AIRLANDBATTLE21

AirLandBattle21

Transformational Concepts for Integrating Twenty-First Century Air and Ground Forces

ELLWOOD P. ("SKIP") HINMAN IV
U.S. Air Force, USA

THOMAS E. ("TOM") JAHN
U.S. Air Force, USA

JAMES G. ("J.J.") JINNETTE
U.S. Air Force, USA

ASHGATE

Published by
Ashgate Publishing Limited
Wey Court East
Union Road
Farnham
Surrey, GU9 7PT
England

Ashgate Publishing Company
Suite 420
101 Cherry Street
Burlington
VT 05401-4405
USA

www.ashgate.com

British Library Cataloguing in Publication Data
Hinman, Ellwood P.
AirLandBattle21 : transformational concepts for integrating twenty-first century air and ground forces
1. Unified operations (Military science) 2. Air power - United States 3. Air warfare - United States 4. Military doctrine - United States
I. Title II. Jahn, Thomas E. III. Jinnette, James G.
355.4'6'0973

Library of Congress Cataloging-in-Publication Data
Hinman, Ellwood P., 1966-
AirLandBattle21 : transformational concepts for integrating twenty-first century air and ground forces / by Ellwood P. Hinman, Thomas E. Jahn, and James G. Jinnette.
p. cm.
Includes bibliographical references and index.
ISBN 978-0-7546-7634-8
1. Unified operations (Military science) 2. Air power--United States. 3. Air warfare--United States. 4. Military doctrine--United States. I. Jahn, Thomas E. II. Jinnette, James G. III. Title. IV. Title: Air Land Battle 21, transformational concepts for integrating 21st century air and ground forces. V. Title: Transformational concepts for integrating twenty-first century air and ground forces.
U260.H55 2008
355.4'60973--dc22

2008038306

ISBN: 978 0 7546 7634 8

Mixed Sources
Product group from well-managed forests and other controlled sources
www.fsc.org Cert no. SGS-COC-2482
© 1996 Forest Stewardship Council
FSC

Printed and bound in Great Britain by
TJ International Ltd, Padstow, Cornwall

Contents

List of Figures

Notes on Contributors

Ellwood P. ("Skip") Hinman IV, Thomas E. ("Tom") Jahn, and James G. ("J.J.") Jinnette are U.S. Air Force pilots with over 9,500 combined hours flying eight different fighter, airlift, and training aircraft. All three officers graduated from the Air Force Academy and have earned seven master's degrees throughout their careers. They have accumulated a total of over 56 years of military service in numerous assignments from the squadron level to the Office of the Secretary of Defense, including tours commanding at the squadron level, developing strategy on a numbered Air Force staff, writing doctrine at the Air Force Doctrine Center, and teaching at Air University. They have collectively participated in several contingency and combat operations to include Desert Storm, Provide Comfort, Southern Watch, Allied Force, Phoenix Scorpion, Enduring Freedom, and Iraqi Freedom.

Disclaimer

The views expressed in this study are those of the authors and do not reflect the official policy or position of the United States Government, Department of Defense, or the United States Air Force.

Chapter 1
Laying the Foundation

> New conditions require, for solution—and new weapons require, for maximum application—new and imaginative methods. Wars are never won in the past.
>
> —General Douglas MacArthur

The changed strategic landscape of the Twenty-first century has driven the Department of Defense (DoD) to direct transformation in the U.S. military. In essence, this mandate requires a shift from legacy technologies and Cold War organizations to more flexible, adaptable capabilities and constructs effective across the spectrum of conflict. At its core, this transformation represents a revolution in military affairs (RMA). Most define an RMA as the confluence of new technologies, new organizations, and new ideas. Each of the Services has embarked on aggressive plans that feature many transformational technologies accompanied by new organizational constructs. The U.S. Army, for example, has shifted to modular forces based on the brigade combat team (BCT) and new technologies such as the Stryker infantry vehicle and many others incorporated into the Future Combat System (FCS).

For the revolution in military affairs to be complete, though, the same transformation must take place in the area of ideas. These new concepts must bridge the ongoing changes within the separate branches and integrate the developing Service-specific technological capabilities into a seamless whole. More specifically, the need exists for a coherent set of transformational concepts that fully integrate Twenty-first century American airpower with the Army's new brigade combat team. The purpose of this study, then, is to offer innovative thought in the areas of counterland, counterair, strategic attack, and tactical airlift that will merge on-going transformational efforts in the Air Force and Army into a single, potent fighting force. In the last century, the threat of war with the Soviet Union in central Europe led to a similar transformation after the U.S. debacle in Vietnam that culminated with the development of an operational concept for air and ground forces labeled *AirLand Battle*. AirLand Battle fell away with the Berlin Wall and a comprehensive approach has yet to replace it. The ideas developed in this book will form the baseline for a novel military strategy toward a new AirLand Battle construct for the transformed context of the Twenty-first century.

The Need for Transformational Ideas

The 2006 *National Security Strategy* observed that "the major institutions of American national security were designed in a different era to meet different challenges. They must be transformed" and proclaimed that "the United States must transform America's national security institutions to meet the challenges and opportunities of the 21st Century" (*National Security Strategy* 2006, 1, 43). While former Defense Secretary Donald Rumsfeld initially led the DoD transformation, his successor, Secretary Robert Gates clearly stated during the November 2006 Senate confirmation process his intent to stay the transformational course: "If confirmed, I will build upon the President's commitment to transform our forces to better fit the 21st century. Transformation holds the promise to ensure that our military forces are more agile and lethal when confronting the enemies of this new century." (Gates November 2006, 11). In his initial message to DoD personnel, the Secretary Gates formalized and reaffirmed his position:

> It is important that we carry forward the process of what has been referred to as defense transformation. President Bush made it a priority at the beginning of his administration to help our military become more agile, more lethal, and more expeditionary. Much has been accomplished; much remains to be done. This task remains a necessity and a priority today (Gates December 2006).

One could argue that what has been accomplished is the advent of new technology and organizational constructs. What remains to be done, then, would be the inclusion of concepts that will complete the transformation.

While the term *transformation* means different things to different people, it is generally accepted that the goal is a smaller, more deployable, high-tech force more responsive to the changed strategic environment of the Twenty-first century. Each of the Services and many of the joint combatant commands have embarked on aggressive plans that feature many transformational technologies accompanied by new organizational constructs. The Army's new baseline fighting force, the BCT, is arguably the most visible and among the most important of these organizational changes. The DoD-mandated transformation has arguably directed a shift from legacy technologies and Cold War organizations to more flexible, adaptable capabilities and constructs which are also more compatible with the new strategic landscape. Indeed, the individual Services and combatant commands are moving in this new direction.

In 2003 guidance to the Department, the Office of the Secretary of Defense defined transformation as "a process that shapes the changing nature of military competition and cooperation through new combinations of concepts, capabilities, people and organizations that exploit our nation's advantages and protect against our asymmetric vulnerabilities to sustain our strategic position, which helps underpin peace and stability in the world" (*Department of Defense Transformation Planning Guidance* April 2003, 3). David Fastabend, a U.S. Army general officer, found that "transformation is most typically associated with the 'revolution in military affairs', that is, the technological changes, organizational changes, and operational concepts we typically associate with information technology, extended range, and speed and precision" (Fastabend 2005, 155-156). Both definitions, then, identify the ingredients of an RMA, namely new technologies, new organizations, and new ideas, as the basis for transformation. Some might chafe at the implication that current defense-related changes amount to an RMA, citing evolutionary—vice revolutionary—adjustments. However, fewer of these skeptics would argue with the assertion that, regardless how one might classify recent these technological and organizational developments, such change in these areas necessitates equally innovative ideas as well.

While technological and organizational innovation continues at a feverish pace within the defense establishment, though, the infusion of new ideas has not kept the same pace. Fastabend went on to rhetorically query "where are the fundamentally new ways to visualize warfare or operations?" and answered by arguing that, "sadly, there are all too few. To date, we generally apply faster drying paint to draw the same type of picture more rapidly and precisely. There is all too little transformation in our operational concepts, particularly at the joint level" (Fastabend 2005, 157). In essence, there appears to be an incremental evolution in military theory while the new strategic environment of the Twenty-first century demands revolutionary change. Despite transformation elsewhere, then, operational concepts for prosecuting warfare within this new context remain wedded to legacy ideas developed decades ago. Although generally accepted in the past century, the practice of "pouring new wine in old bottles" may no longer suffice. Accordingly, a fundamental argument in this paper is that, while the U.S. military is leaning forward with transformational change in the areas of technology and organization, the requisite development of new ideas that provide the operational framework for such change has been slower. For the RMA to be complete, the same transformation must take place in the area of ideas. This study, then, aims to answer Fastabend's admonition with transformational thought in the same joint arena where he saw it sorely lacking.

Regrettably, limits on the development of transformational concepts might be related to an emphasis on technology. In an essay titled "The American Way of War", Colin Gray found the United States defense establishment obsessed with technology, specifically relating that fixation both to American airpower and the nation's ground forces:

> The exploitation of machinery is the American way of war. One may claim that airpower is virtually synonymous with the American way of war and that its employment as the leading military instrument of choice has become routine … America is the land of technological marvels and of extraordinary technology dependency … American soldiers say that the human being matters most, but in practice the American way of war, past, present, and prospectively future, is quintessentially and uniquely technologically dependent (Gray 2005, 29).

In a later chapter of the same book, Grant Hammond submitted a similar argument. He began by citing former Air Force Colonel John Boyd's dictum, "people first; ideas second; things third", and then showed America's divergent approach:

> Unfortunately, the U.S. military tends to reverse the priorities. Oddly enough, that is in part because of a concern for people. Rather than masses of manpower to accomplish a military mission, the United States seeks to put fewer people in harm's way, to substitute technology and firepower for manpower. The emphasis on technological solutions for problems is a peculiarly American trait. Nowhere is that more apparent in the high technology investments made in air, space, and information power (Hammond 2005, 123-124).

While today's military emphasizes hardware, then, there tends to be less emphasis on what might be called the "software" that brings the technology to life. While problematic at any period in American history, this phenomenon would be even more troublesome in the Twenty-first century. As Robert Scales argued, "more than ever war is a thinking game. Wars today must be fought with intellect as well as technology" (Scales 2005, 43). Similarly, former Army Chief of Staff General Peter Schoomaker made the critical point that "transformation is not about equipment. It's about intellect; it's about judgment; it's about the development of leaders and soldiers. You've got to make that intellectual transformation before you can make the visible transformation" (Risher 2005, 245).

In his forward to the 2004 *National Military Strategy*, Chairman of the Joint Chiefs of Staff, General Richard B. Myers, emphasized the importance of not just technological development but also concept development as the military transforms. He explained that transformation involved both "fielding new capabilities and adopting new operational concepts." He went on to submit that "transformation requires a combination of technology, intellect and cultural adjustments—adjustments that reward innovation and creativity" (*National Military Strategy* 2004, v). Later in the same document, writers drove home the point:

> While the United States enjoys an overwhelming qualitative advantage today, sustaining and increasing this advantage will require transformation—a transformation achieved by combining technology, intellect and cultural change across the joint community. The Armed Forces must be able to evaluate challenges, leverage innovation and technology and act decisively in pursuit of national goals (*National Military Strategy* 2004, 15).

Successfully completing the defense transformation, then, may require a parallel effort toward cultural transformation. Such an effort might shift the American military's past cultural fascination with technology toward an equitable emphasis on the development of innovative ideas.

Conventional War Remains Relevant

In addition to the need for transformational concepts, this study was also inspired by the need to refocus America's cognitive effort on medium- to large-scale conventional war. Much ink has been spilled of late on unconventional war, international terrorism, failed states, transnational weapons of mass destruction, conflict with non-state actors, counterinsurgency operations, and similar trends dubbed *fourth-generation warfare* (4GW). The Services, the joint community, and the Defense Department as a whole are spending substantial collective time and intellectual energy addressing such issues. The 2006 *Quadrennial Defense Review* (*QDR*) explained the transformation as "a shift of emphasis to meet the new strategic environment." One quite pertinent example it highlighted was a shift "from major conventional combat operations—to multiple irregular, asymmetric operations" (*Quadrennial Defense Review* 2006, iv). A wholesale shift in this new direction, however, could have dangerous implications for the U.S. Armed Forces and their ability to fight conventional war in the future.

The 2008 *National Defense Strategy* (*NDS*) asserted that,

> U.S. predominance in traditional warfare is not unchallenged, but is sustainable for the medium term given current trends. The 2006 QDR focused on non-traditional or irregular challenges. We will continue to focus our investments on building capabilities to address these other challenges, while examining areas where we can assume greater risk … An underlying assumption in our understanding of the strategic environment is that the predominant near-term challenges to the United States will come from state and non-state actors using irregular and catastrophic capabilities (*National Defense Strategy* 2008, 21-22).

For the most part, this assumption may very well be true. Unfortunately, it may lead some in DoD to expend disproportionate effort on irregular methods at the expense of a very real traditional threat. A key point not recognized by these documents is that, if American brainpower swings too far for too long in this new direction, its potential adversaries may perceive a new "asymmetric" advantage in atrophied and outdated U.S. traditional capabilities. In fact, a report released by House Armed Services Committee in December 2006 cited this possibility: "Despite the increased demands of the GWOT [global war on terror], the committee does not believe that the requirement to be prepared to fight and win two major regional contingencies simultaneously has been rendered moot. In fact, the committee believes that the requirement may have increased as potential aggressors see opportunities to exploit the U.S. commitment to the GWOT" (*Committee Defense Review Report* 2006, 70).

Irregular warfare is a critically important dimension of war in this century and one that was largely foreign to the prevailing concepts for conflict in the last century. America must meet the challenges of the Long War with continued effort in this area. However, what once was the niche is now the mainstream. Out of favor at present is focused study on medium- to large-scale conventional conflict against a state with credible ground and air forces. It is certainly true that the war on terror and the current conflicts in Iraq and Afghanistan mandate the development of new ideas in the area of irregular war. However, as Air Force Lieutenant General David Deptula argued, "it's not enough to fight today's war against today's enemy. We must be prepared for tomorrow" (Deptula 2007, A14).

Non-traditional conflict or 4GW is certainly *a* likely form of war in the future but, contrary to the teachings of some contemporary thinkers, it is by no means *the* future of war. It would be quite problematic and arguably dangerous to shift too far in this new direction. Current focus on the Long War and the war in Iraq may be blurring what should be a very clearly evident conventional threat to America's national security in the Twenty-first century. The American defense establishment also needs novel concepts for "regular war" in a new era of increasingly modern technology. Conventional war against an enemy with a credible land and air force is equally important and likely in the future. The Congressional report concluded that "state-on-state conflict remains, and will be, a major component of the international system. The threat panel identified several countries with significant military capabilities that were developing in a direction that could challenge the U.S. Armed Forces: the People's Republic of China (PRC); India; Pakistan; North Korea; Iran; and Russia" (*Committee Defense Review Report* 2006, 20). The 2008 *NDS* also acknowledged the potential for competition from Iran, North Korea, China, and Russia (*NDS* 2008, 3-4, 13). In fact, just following the June 2008 publication of this document, Russian aggression toward Georgia in August of the same year offered evidence of conventional threats to the United States and its partners around the globe.

Both the Army and Air Force recognize the continued relevance of conventional war in the Twenty-first century. Retired Air Force General Ronald E. Keys, former commander of the Air Force's largest major command, Air Combat Command, argued that national security considerations go beyond Iraq and Afghanistan: "You've got to be able to fight North Korea … You've got to be able to defend in the China–Taiwan Strait. You've got to be able to go to Iran" (Dudney 2006, 2). Robert S. Dudney, editor of *Air Force Magazine*, added that "such scenarios would entail high-intensity clashes with large national forces. Those nations could be defeated only by a technologically advanced 'conventional' military" (Dudney 2006, 2). Renowned Air Force historian Phillip Meilinger agreed:

> Airmen must make it clear that traditional threats—a major conventional war against a peer competitor—have not disappeared. No one predicted twenty years ago that the Cold War would soon be over and that the U.S. would then fight *two* major wars against Iraq—a country we had been courting as a balance against a volatile Iran—while also conducting two other major air campaigns in the Balkans (Meilinger, unpublished manuscript, 31, emphasis in original).

A 2006 Army document that analyzed the strategic environment in the next few decades arrived at conclusions remarkably similar to their sister Service:

> Conventional operations conducted within a state-on-state framework will continue to be relevant in the future environment. States will remain wedded to strategies based on the use of military power to achieve their goals, in conflicts that range in size from smaller scale contingencies (SSC) to MCO [major combat operations] and occur in often unforeseen locations and varied climatic and topographical conditions. As recent events have shown, even smaller scale or stability operations may escalate with little warning into larger scale hostilities. Thus, regional aggressors will continue to modernize conventional forces and invest in capabilities that will enable them to dominate their neighbors (*The United States Army's Operating Concept for Operational Maneuver: 2015-2024* 2006, 5).

The 2006 *National Security Strategy* explained that DoD is "transforming itself to better balance its capabilities across four categories of challenges." In addition to catastrophic and disruptive challenges, the *National Security Strategy* acknowledged "irregular challenges from state and non-state actors employing methods such as terrorism and insurgency to counter our traditional military advantages." The fourth category, most relevantly, involved "traditional challenges posed by states employing conventional armies, navies, and air forces in well-established forms of military competition" (*National Security Strategy* 2006, 43-44). Perhaps most important in this *National Security Strategy* argument was the word *balance*. One could argue that the recent lopsided emphasis on irregular threats such as terrorism and insurgency has tipped the scale. The current focus on Iraq and the war on terror threatens to upset this necessary balance. It invites disproportionate focus on 4GW at the expense of other more traditional, conventional threats posed by nations such as North Korea, China, Iran and Syria. As Robert Dudney observed, "some new emphasis on irregular threats was warranted, but overcorrection can be dangerous, and it is not easy to know when that has happened" (Dudney, 2). Accordingly, as he prepared to take charge of DoD, Secretary Robert Gates seemed to indicate that transformation under his leadership would address conventional threats posed by states along with others in a balanced fashion: "Transforming the Department to better deal with 21st-century challenges, a major charge from the President, must continue. These challenges range from the threat posed by terrorist networks, to weapons of mass destruction in the hands of hostile regimes and terrorist networks, to states armed with advanced weapons" (Gates, November 2006, 8).

The essence of transformation has been a shift from a U.S. military optimized for one war—a large, conventional war—to one fully prepared to fight more flexibly across the spectrum of conflict. This has been both necessary and appropriate. The risk now is that the pendulum seems to be swinging too far to the other side. Too often, militaries fall victim to the temptation of "fighting the last war." Because the United States faces future threats across the spectrum of conflict, it is critical that strategic thinkers avoid becoming too narrowly focused on the current struggle. This study aims to swing the pendulum in the other direction by paying increased attention to major combat operations. The goal is a coherent body of concepts for medium- to large-scale conventional conflict. While the ideas that follow may have applicability to small wars and irregular warfare, they are designed for wars between the U.S. military and states possessing credible air and ground forces.

A New Era of Jointness

Joint Publication (JP) 1-02, the *DoD Dictionary of Military and Associated Terms*, asserts quite generically that the term *joint* "connotes activities, operations, organizations, etc., in which elements of two or more Military Departments participate" (*DoD Dictionary*). As the definition suggests, efforts in the last century to achieve jointness often simply enabled Service components to successfully participate in the same operations. To be fair, the last decade of the Twentieth century saw great strides in the development of jointness within the combatant commands. One could argue that these organizations are truly joint while the individual Services themselves are still making the journey. Jointness, then, has matured within the combatant commands but remains adolescent within the Services. These then are the formative years for true inter-Service jointness.

Jointness in the holistic sense is the way in which wars are fought, not simply the way to organize warfighting forces. Many forward-thinking concepts are currently developed within the Services, rather than the joint community, and are currently largely Service-oriented. Efforts to cross Service lines and move into the joint realm have focused primarily on deconfliction between individual Service concepts and capabilities and have traditionally been less successful beyond this first step. Just as DoD met with great success in developing joint organizations in the last century, the U.S. military needs to take jointness to the next level in this century with operational concepts and ideas for fighting across Service boundaries in a truly joint fashion. Joint concepts supporting this advanced construct would explore how these forces truly fight together as co-equal entities rather than simply next to each other with one in support of the other.

If jointness was the word for inter-Service cooperation in the past, "interdependence" more accurately captures this relationship in the Twenty-first century. There is a subtle but important distinction between these two terms. Jointness reflects the desire for Services to work together. With interdependence, working together is no longer an option but rather a necessity. Jointness of old implied the goal of interoperability between Services and their capabilities. According to retired Army General Gary Luck, however, "interdependence is more than interoperability—the technical ability to work together. It is the recognition that the Armed Forces fight as one team of joint, interagency, and multinational partners—and depend on access to each other's capabilities to succeed" (Luck, 1-37).

In the past, robust capabilities within each Service and redundancy between them allowed for a certain level of Service autonomy, giving them the option of working together or functioning separately. However, what was once desirable will soon be required. In its 2006 forward-looking document on operational maneuver, the Army fully acknowledged this dynamic:

> Joint operational experience over the past 20 years shows steady progress towards achieving joint interoperability, but the Services must become more interdependent to fully exploit and optimize the capabilities of each component. In fact, neither the individual components nor the joint force overall can fully accomplish assigned missions without the participation and contribution of all components. Although each Service contributes its own unique capabilities to the joint campaign, each dominating its own environment, their operational and even tactical interdependence will be critical to overall joint force effectiveness. Joint interdependence is achieved through the deliberate reliance of each Service on the capabilities of others to maximize its own effectiveness while minimizing its vulnerabilities (*The United States Army's Operating Concept for Operational Maneuver: 2015-2024* 2006, 15-16).

Deconfliction and joint participation may have been sufficient for success in the past. In the future, however, the Services will literally and intentionally depend on each other in order to achieve the same results. Another visionary 2006 Army publication found that interdependence "encompasses the central idea of deliberate dependence by ground forces on the employment of joint capabilities, in lieu of organic capabilities, to help accomplish tactical and operational objectives" (*The United States Army Concept for Tactical Maneuver for the Future Modular Force: 2015-2024* 2006). Simply stated, then, interdependence takes jointness to the next level.

To a large extent, this idea of interdependence has come as the result of a fiscally constrained environment where limits on individual Service budgets has and will continue to force mutual dependence on the separate branches. In short, less funding and fewer resources translate to increased interdependence. Specifically, a smaller pocket book, paired with the bigger bills from transformation, has driven the Army, as an example, to reduce certain organic capabilities. In so doing, Soldiers have become increasingly dependent on Airmen in certain mission areas. All told, much is being said about interdependence; less is being done. A central theme in this study, therefore, goes beyond simply recognizing the existence of interdependence by developing practical ways and transformational concepts allowing air and ground forces to operate in this increasingly interdependent environment.

Jointness as practiced in the last century suggested a level of coordination between the Services consistent with *synchronization*, a term commonly used in the U.S. Army. The *DoD Dictionary* defines synchronization as "the arrangement of military actions in time, space, and purpose to produce maximum relative combat power at a decisive place and time" (*DoD Dictionary*). While this joint definition accurately captures the Army's connotation of the word, the word as defined in Webster's more literally—and more simply—refers to the act or result of operating at the same time. Jointness, as previously mentioned, denotes two Services participating in the same operation and synchronization similarly suggests that those forces simply operate at the same time.

Other terms may be necessary to fully grasp the essence of interdependence. The term *integration* may be up to the task. Joint publications define integration as "the arrangement of military forces and their actions to create a force that operates by engaging as a whole" (*DoD Dictionary*). Synchronization may have been appropriate in the past and may still be applicable to coordination between ground forces. Integration, however, seems more compatible with current and future trends and particularly more suited to the emerging relationship between air and ground forces. Accordingly, the 2003 *Air Force Basic Doctrine* document drew a clear distinction between the two terms:

> Synchronization is, in essence, deconfliction in time and space between different units. It is a useful means to plan and execute operations and to prevent fratricide … However, it doesn't scale up to the operational level and hence is not the best means for achieving the maximum potential of a joint force. Synchronization emphasizes timing, while integration considers priority and effect to be both efficient and effective with scarce resources. Synchronization is bottom-up; integration, on the other hand, starts at the top with a single cohesive plan and works downward. Synchronization is an additive "sum of the parts" model, while integration may produce geometric results. This is not to say that synchronization is bad. For surface forces, it is very useful for managing their scheme of maneuver. However, from an airman's perspective, synchronization is a tactical tool and doesn't necessarily scale up to the operational level (*Air Force Basic Doctrine* 2003, 6).

Rather than two separate entities simply participating in a joint operation at the same time, then, integrated air and ground forces would operate as a single entity. The emerging concepts of interdependence and integration imply that air and ground forces are no longer two separate coins being spent on the same military operation but rather two sides of the same coin. The synergy that may be achieved through this new relationship and the holistic application that would follow suggest that this new coin may be far greater in value then the sum of the two previous ones.

The 2006 *Quadrennial Defense Review* explained the transformation as shifts of emphasis "from separate military Service concepts of operation—to joint and combined operations" and "from forces that need to deconflict—to integrated, interdependent forces" (*Quadrennial Defense Review*, vi-vii). It went on the state that "the complex strategic environment of the 21st century demands greater integration of forces, organizations and processes" (*Quadrennial Defense Review*, 4). *Air Force Basic Doctrine* rightfully observed that integrating the capabilities of different Services in this way goes beyond the acquisition of advanced technology: "Effective integration involves more than smart technology investment—it also requires investigation of efficient joint and Service organization, and innovative operational thinking" (*Air Force Basic Doctrine*, 75). As with transformation, then, full integration requires new organizations, in order to facilitate improved command and control, and new ideas as well. In a 2003 *Field Artillery* article titled "Transforming Joint Air-Ground Operations for 21st Century Battlespace", Air Force Major General David Deptula and Lieutenant Colonel Sigfried Dahl found that "the need to rethink how the Air Force and Army synthesize transformation initiatives to best facilitate victory in the JAGO [joint air-ground operations] environment is paramount" (Deptula and Dahl 2003, 280). Similarly, the 2004 *National Military Strategy* stated that "to succeed, the Armed Forces must integrate Service capabilities in new and innovative" ways (*National Military Strategy* 2004, 27). Such is the purpose of this study.

Beyond AirLand Battle of the Cold War Era

In the aftermath of the Vietnam War, the U.S. Army reinvented the way it planned to fight and ultimately codified this new approach in the 1982 publication of *Field Manual (FM) 100-5*. The novel combined-arms doctrine encapsulated in this document was titled *AirLand Battle*. While officially applicable to any conflict across the globe, most observers viewed AirLand Battle as the new strategy for fighting numerically superior Soviet forces in Western Europe. The new doctrine incorporated many new and refined ideas. The 1982 Army field manual clearly defined and codified for the first time in U.S. military doctrine the operational level of war (*FM 100-5* 1982, 2-3; Long 1991, 64-65). It broke with previous doctrinal assumptions of a wholly linear battlefield, arguing that "opposing forces will rarely fight along orderly, distinct lines." Instead, *FM 100-5* found that massive troop concentrations or immensely destructive fires will make some penetrations by both combatants nearly inevitable. This means that linear warfare will most often be a temporary condition at best and that distinctions between rear and forward areas will be blurred (*FM 100-5* 1982, 1-1 to 1-2; Long 1991, 63-64). While these precepts were indeed new to U.S. military doctrine, the most notable, fundamental and novel aspect of AirLand Battle was the concept of the deep battle.

General Donn Starry, the Commander of the U.S. Army's Training and Doctrine Command (TRADOC) who is generally credited with the creation of AirLand Battle, served as V Corps Commander in Europe prior to his post at TRADOC. While in Europe, Starry identified fundamental flaws in the defensively-oriented approach, entitled *Active Defense*, which preceded AirLand Battle in the 1976 publication of *FM 100-5* (Romjue 1993, 55). In focusing primarily on the initial close battle, he found that the Active Defense doctrine failed to address the hordes of second-echelon Soviet forces that would reinforce the front-line units engaged in the initial battle in the Fulda Gap (Murray and Scales 2003, 46-47). Starry's solution to this conundrum was to view the battlefield more holistically. Rather than focus myopically on the front lines, he widened the aperture by also attacking enemy units not yet engaged in the initial ground battle. In accordance with his novel approach, forces that could reach the enemy's second-echelon would attack deep while North Atlantic Treaty Organization (NATO) forces fought the close battle. Because the U.S. Army did not yet have artillery systems with the necessary range, the critical responsibility for fighting the deep battle fell almost exclusively to NATO airpower.

For Services that were not accustomed to fighting as an integrated team, AirLand Battle was indeed a watershed concept. It has been described as a "doctrinal phenomenon", "a renaissance in doctrinal thinking", and "a unique concept" with "revolutionary ideas" (Romjue 1993, 51; Leonhard 1991, 129). In their book, *The Iraq War: A Military History*, Williamson Murray and Robert Scales discussed the associated transformation in the U.S. Armed Forces that followed the debilitating impact of the Vietnam War: "The story of that dramatic comeback is one of the greatest in American military history" (Murray and Scales 2003, 46). They cited "the reinvention of the military as a serious profession—one that demanded intellectual as well as physical prowess. For officers, the latter challenge required serious focus on the complex tasks involved in articulating violence on the battlefield" (Murray and Scales 2003, 46). It is important to note that transformation in the last century featured not just technological advances and organizational change, but placed highest priority on the need for transformational thought. According to Murray and Scales, it was this mindset that led Starry and others in this decidedly new direction (Murray and Scales 2003, 46-47).

AirLand Battle of old, however, was not without its detractors, especially in the U.S. Air Force. As one Air Force official explained, "when we say we agree with the AirLand Battle concept … we agree that the concept is a good concept *for the Army*" (Gordon 1983, 4F; London 1984, 26, emphasis added). Fundamental to this nuance was that, while commendable that the Army uniquely recognized a critical role for airpower, it once again relegated the Air Force to its traditional supporting role. In a brief section characteristically titled *Air Support, FM 100-5* submitted that "the Air Force is an equal partner in the air-land battle" but then went on to say that "it supports the battle" with a list of Air Force missions intended to be exhaustive (*FM 100-5* 192, 7-6 to 7-7). The field manual similarly specified that "the basic combined arms maneuver element is the battalion task force" and stated unequivocally that "Air Force and Army air elements provide support" (*FM 100-5* 1982, 7-4). It went on to articulate airpower's most central contributions to AirLand Battle and defined its missions in sections titled *Fire Support* and *Offensive Air Support* (*FM 100-5* 1982, 7-10 to 7-11). AirLand Battle, then, blazed new trails in U.S. military combined-arms strategy but left other elements of true jointness and integrated air-ground operations largely unexplored.

Similarly, the 1982 Army field manual was silent for the most part on the many critical roles airpower would play in a war with the Soviet Union and in other conflicts around the world. For example, while formally co-equal, there was no chapter or large section of *FM 100-5* dedicated to airpower's key contributions. One could argue quite convincingly that, beyond specific references to airpower totaling less than one page in over 110 pages of text, there was little "air" in the initial AirLand Battle concept (FM 100-5 1982, 7-6 to 7-7, 7-11, 7-13). Notably absent was any substantive discussion of air superiority, tactical airlift, or ground attack roles beyond the close battle and the enemy's second echelon.

Finally, this support-centric, somewhat shortsighted approach limited the potential efficacy of air-ground operations. Robert R. Leonhard explained in *The Art of Maneuver: Maneuver-Warfare Theory and AirLand Battle* that air attacks envisioned in AirLand Battle of old present the enemy "with a problem, but not a dilemma." By narrowly placing airpower exclusively in direct support to ground forces, the doctrine did not opt for a multifaceted approach that could force the enemy's hand. He argued that "deep operations will continue to be one dimensional and easily countered under current doctrine. The unplayed trump card is joint (ground and air) operations" (Leonhard 1991, 163). The dilemma absent in the Army's Cold War combined-arms concept is central to the truly integrated air and ground operations advocated in *AirLandBattle21*.

AirLand Battle is a term that incites much passion. For airpower enthusiasts familiar with the concept, that sentiment most often comes in the form of anger. Others who view warfare from the Soldier's perspective generally demonstrate great fondness for the Cold War strategy. This book aims to bring the familiar phrase into the Twenty-first century and reinvigorate the debate without taking sides. Instead, the intent is to refocus attention on an area underemphasized since AirLand Battle fell away in the aftermath of the Cold War. While passion remains on both sides, *AirLandBattle21* aims to bridge the divide and build a single team by integrating air and ground forces in ways unachieved in the last century.

AirLand Battle was the last comprehensive U.S. military strategy that specifically addressed how air and ground forces would fight a major conventional war. The Soviet Union no longer exists, and the Fulda Gap is little more than a geographic feature interior to a unified Germany. The Twentieth century had AirLand Battle; no such coherent approach exists in the new century. Consequently, the time has come for new concepts integrating Twenty-first century air and ground forces that will fight interdependently on the battlefields of the new millennium. This is the purpose of *AirLandBattle21*.

Many would argue that the mechanism for providing such innovative thought is joint doctrine. Some might even contend that the body of joint doctrine publications has replaced AirLand Battle of old. On the contrary, joint doctrine is written by consensus and consequently tends at times to be less responsive to dynamic changes such as those currently taking place with technology and organizations. Air Force Major Mark Czelusta recognized this dynamic in 2003 and argued "the perhaps greatest weakness in U.S. joint doctrine" is that it falls short of full inter-Service integration:

> While clearly supporting the concept that integration is the ultimate goal of operations, it offers few techniques to integrate below the JFC and his staff. Below this level, emphasis is on coordination and liaison, vice integration. As such, current doctrine devotes significant discussion to methods for division of the battlespace, and placement of liaison teams within each component's command organizations. If integration is to be successful, additional emphasis on how to be successful in the same battlespace, vice how to coordinate activities across areas of operation, is required. Another significant weakness is doctrinal responsiveness to changes in technology and capability. This weakness is especially evident as one tries to incorporate doctrine related to specific capabilities and tasks to a larger operation … Each joint doctrine manual discusses phases of the respective operation or task, but offers precious few techniques for successful simultaneous activities (Czelusta 2003, 16-17).

Along the same lines, the 2003 *Air Force Basic Doctrine* document acknowledged the limitations of both Service and joint doctrine in their efforts to achieve full integration and agreement in developing areas:

> In circumstances when the Air Force cannot find a unanimous doctrinal consensus, it may settle on an "agreed-to, least-common-denominator" position that all players are willing to sign up to. This frequently occurs in emerging mission areas, where new concepts and terms have yet to solidify across the Service. It may also occur within joint doctrine when two or more Services propose conflicting changes to joint doctrine; to avoid deadlock, and to keep the joint doctrine development process in motion, the Services may consent to "agreed-to" language (*Air Force Basic Doctrine* 2003, 86).

Air Force integration with the new Army BCT construct is arguably one such emerging area. As such, novel ideas within the two Services are forthcoming, but one can not yet find fully developed transformational concepts that transcend a single branch, cross Service lines, merge these new concepts into a single comprehensive theory, and integrate Twenty-first century forces into a seamless fighting force. This study will provide this thinking unconstrained by the inherently-contentious joint-doctrine process.

Retired Army Lieutenant Colonel Robert Leonhard found that AirLand Battle of the Cold War era "provided the foundation for a radical change in how the army (and by extension, the joint community) fought" (Leonhard 2005, 209, 215). Two important points become clear from this assertion. The first is that AirLand Battle embodied transformational concepts. Second, while envisioning airpower's contribution to the battle in new ways, AirLand Battle addressed joint warfare from a decidedly Army viewpoint. AirLand Battle was in fact designed by an Army general and was accordingly conceived from the Soldier's perspective. As a result, while AirLand Battle of the Cold War era was generally viewed favorably across the military and certainly in the U.S. Army, some Airmen held negative views.

Many in the Air Force found the military strategy to be decidedly ground-centric in its view of the battlefield. Simply stated, AirLand Battle of old was, for many Airmen, Army doctrine. Because the Air Force doctrine process was less developed at that time and joint doctrine did not exist, it could be said that AirLand Battle filled the void. The prevailing Air Force view holds that the Cold War concept generally placed the Air Force in support of the Army scheme of maneuver in the traditional and all-too familiar role of "winged artillery."

The goal of *AirLandBattle21* is to offer transformational concepts about joint warfare as did its predecessor but to do so with a more balanced approach built upon viewpoints from both the Army and the Air Force. Because the authors of this study are Air Force officers, some might argue that the ideas embodied in *AirLandBattle21* have been developed from the Airman's perspective. On the contrary, every attempt has been made to fairly represent the Soldier's vantage point on the ground as well as the Airman's perspective from above. The authors conducted multiple surveys and interviews with field-grade Army officers, made countless communications with Army officers and professors at several war colleges, and researched the most current pertinent Army doctrine and publications. Some readers may label portions of this text as airpower theory. If an accurate appraisal, it is one well grounded, quite literally, in the new BCT construct. It looks up from the ground as well as down from above, with conclusions decidedly different from the Airman's traditional approach. In sum, if AirLand Battle was the first big step toward Army-Air Force jointness, then *AirLandBattle21* will be an equally large step toward a set of novel concepts fully integrating and enabling interdependence between the two Services.

Bounding the Focus

As previously discussed, the intent of this study is to isolate and analyze the interaction between U.S. Army ground forces and U.S. Air Force airpower. The context for this interaction is medium- to large-scale conventional war. As such, the ideas will have some but often limited applicability to conflicts of an irregular nature. Furthermore, while space and information operations are critical in today's and tomorrow's wars, this study does not specifically address these areas in a comprehensive manner. *AirLandBattle21* aims to unite the expeditionary Army and the Expeditionary Air Force into a single fighting force. This union will in some ways resemble the Marine Air-Ground Task Force (MAGTF) applied at the operational level of war. Interestingly, the Marine Corps is also moving toward distributed operations with brigade and regimental combat teams. Additionally, both the Marine Corps and the Navy bring important airpower capabilities to the joint fight. As such, many of the concepts discussed in the following pages would certainly be applicable to a joint fight that includes maritime forces. Nonetheless, this study is focused squarely on the Air Force and Army and will not specifically address Marine Corps air and ground forces or Navy air and maritime forces in any substantive way. Similarly, the interagency and multinational aspects of modern warfare, of greatly increased importance in the Twenty-first century, receive no direct attention in the pages that follow.

Finally, *AirLandBattle21* will not offer a panacea or a *one-size-fits-all* approach to future wars. It is unlikely every idea in the pages that follow would perfectly fit any future operation. While it addresses a fairly inclusive list of airpower mission areas, *AirLandBatte21* would hardly constitute a fully complete strategy for a particular military operation. Future planners, however, may find the menu of options appetizing. Armed with only general ideas from joint doctrine and the dated concepts of the last century, these planners would otherwise feel compelled to start from scratch. Instead, the goal of this study would be to afford them with an initial conceptual framework far more compatible with Army transformation and conflict in the changed strategic environment of the Twenty-first century.

In 2006, Marine Corps Lieutenant Colonel Ted Anders opined in a personal communication with the authors that, in current times, "our intellectual tempo is slower than our operational tempo." With the extraordinary time and effort being expended over the wars in Iraq and Afghanistan, perhaps the pace of military thought is slower because of the unrelenting operational tempo over the past five years. When prioritized properly, however, intellectual energy is spent before the action takes place. For ideas to be successful, reasoned development must precede their application. Currently, however, some in the defense establishment have fallen behind the feverish pace of the Long War and are reacting rather than anticipating change. *AirLandBattle21* endeavors to break this chain by increasing intellectual tempo in a highly significant and arguably neglected field of study.

Chapter 2

The Brigade Combat Team Construct

> Today, we are on a path to modernize and transform the Total Army into a modular force that is fully trained, manned, equipped, and supported in a manner that will enable sustained operations in theaters of operation like Iraq and Afghanistan, and those that loom on the horizon.
>
> —General Peter Schoomaker (2006)

In 2001, Williamson Murray wrote that, "at the start of the 21st century, the Army confronts the reality that the ending of the Cold War a decade ago so altered the strategic landscape that virtually all of the strategic verities of a 50-year period disappeared. The adjustment of that reality is still going on." Despite this phenomenon, he observed that, "with the exception of the 7-year period during the Vietnam War, from 1950 on, the Army focused almost single-mindedly on how best to deter and, if deterrence failed, fight a war in Central Europe" (Murray 2001, 2). The intellectual outgrowth of this past effort, of course, was the advent of AirLand Battle. As Soldiers and Airmen alike look to fill the void with a new combined-arms strategy for war in the Twenty-first century, the answer may be *AirLandBattle21*.

The Army's on-going effort toward Twenty-first century transformation aims to develop a lighter and more agile force, potent across the spectrum of conflict. While the Twentieth century Army deployed entire divisions in the event of war, the brigade will become the new baseline combat fighting force. Instead of the months it took to move bulky Cold War divisions, the goal for deploying fully combat-capable new BCTs will be measured in days. Additionally, the new modularity concept will allow one brigade to fall in on another unit's equipment during an in-theater rotation, drastically reducing rotation and reset timelines and associated intertheater airlift requirements. In the last century, the Army was generally more focused on and prepared for the large, conventional war with the Soviet Union. In this century, the goal is a *pentathlon* Army adept at multiple events such as peacekeeping, stabilization, and, most pertinent to this study, major combat operations in any theater across the globe.

The Army's shift to modular forces is part and parcel of the ongoing DoD transformation. A 2002 RAND report traces the roots of Army rebirth farther back to two U.S. military operations in the 1990s, Operation Desert Shield and Operation Allied Force. First, the report explained that, "during Operation Desert Shield, the United States sought to rapidly move sufficient force to the Persian Gulf to defend Saudi Arabia from Iraqi armored forces poised on the Saudi-Kuwaiti border. No Army force at the time was both sufficiently light to move rapidly by air and in possession of the lethality, survivability, and mobility to stop Iraqi armor" (Vick et al. 2002, 2). The 82nd Airborne Division was selected as a stop-gap measure, but later studies suggested that this light division, albeit highly skilled and capable, would not have been able to stop heavy Iraqi units had Saddam Hussein made a move on the Saudi oil fields. Allied Force revealed a similar limitation to Army capabilities: "U.S. planners confronted a wide gap between an air-only effort, which would start almost immediately, and an air-ground effort, which would take months to prepare" (Vick et al. 2002, 2). After much consternation, Task Force Hawk was ultimately deployed but did not see action in the 78-day military operation against Serbia (Vick et al. 2002, 2). The realization of these challenges led then Army Chief of Staff, General Eric Shinseki, to initiate his Service's journey toward transformation. In an October 2000 *Army* article, Shinseki articulated his vision:

> Our legacy Army's warfighting prowess today is assembled around two force characteristics—*heavy* and *light*: magnificent *heavy* forces that are well equipped for war but difficult to deploy strategically, and magnificent *light* forces that can respond rapidly and are well suited for stability and support operation but lack staying power against heavy mechanized forces … With each passing year, our condition as a force becomes a greater liability (Shinseki 2000, 23, emphasis in original).

As a result, the Army embarked on a bold course to fill this gap with modular forces based on the brigade combat team. This transformation continues. During Congressional testimony in June 2006, Shinseki's replacement, General Peter Schoomaker, explained that the Army was "transforming to become a more powerful, more flexible, and more deployable force" (Schoomaker 2006, 4). Ultimately, the Army will field 70 BCTs, with 42 in the active force and 28 in the Guard and reserves. Nineteen will be heavy brigades, armed with Abrams tanks and Bradley fighting vehicles. Forty-four brigades in the transformed Army will be light, infantry units. While each of these BCTs will be far more rapidly deployable than the larger, legacy divisions that preceded them, seven Stryker brigades, equipped with the medium, mobile and survivable Stryker infantry carrier vehicle, will bridge the gap between heavy and light. These Stryker brigades will enable more seamless operations across the spectrum of conflict. Featuring both the new Stryker fighting vehicle and the Army's new BCT construct, the Stryker brigade is the consummate example of new technology and new organizations ripe for new ideas to complete the transformation. The next step in the Army's transformation, called the Objective Force, will integrate the Future Combat System (FCS) into these brigades. With units on three-year steady-state training and deployment cycles, 18-20 BCTs—14 active and four reserve—will be on call to deploy in support of military operations at any given time ("Modular Force Update" 2006).

Distributed Operations

Figure 2.1 shows the Army's emerging concept for operational maneuver of BCTs in the 2015-2024 timeframe. It depicts a clearly non-linear battlefield and displays autonomous BCTs deployed in a non-contiguous and geographically separated fashion. These BCTs will be employed suddenly in the enemy's rear area or perhaps on his flank in order to engage these enemy forces or, according to the Army, to directly attack centers of gravity. The Service's forward-looking document provided a detailed explanation of the new concept:

> In accordance with the joint campaign plan and the integrated actions of other components, the future Modular Force will seize the initiative through shaping and entry operations. Intratheater operational maneuver by ground, air, and sea of powerful, modular, combined arms formations extends the reach of the JFC, enables the force to exploit opportunity, and generates dislocating and disintegrating effects through the direct engagement of decisive points and centers of gravity. Empowered by improved intelligence that contributes to higher levels of situational understanding, the future Modular Force conducts simultaneous, distributed operations within a non-contiguous battlefield

Figure 2.1 The Army's operating concept for operational maneuver

Source: TRADOC Pamphlet 525-3-1, reproduced with permission.

> framework to act throughout the joint operations area (JOA) and threaten the entirety of enemy dispositions, thus hastening the achievement of operational objectives. Continuous operations and controlled (high) operational tempo similarly help overwhelm the enemy's ability to respond effectively and support a pace of physical destruction and psychological exhaustion not achievable today (*The United States Army's Operating Concept for Operational Maneuver: 2015-2024* 2006, iii).

Army officers have varying views on future methods of employing BCTs. Army Colonel Hank St-Pierre, who served as one of the original designers of the BCT construct, explained that the concept was to develop a fighting force "not as powerful but built like divisions at the brigade level" (St-Pierre 2006). In order to achieve this vision, Army planners "took the division construct—an organization that has all the requirements for land combat organic to it—and we brought it down to the brigade. We basically built 'mini divisions'" (St-Pierre 2006). According to St-Pierre, whether or not BCTs are employed in accordance with a distributed operations model completely "depends on the situation. In a situation where the enemy's disposition is not fully known, the brigade commander may own lots of space without any bodies manning it. If you're preparing for a deliberate, preplanned attack, however, the battlefield would be more linear" (St-Pierre 2006). Similarly, Army Colonel Craig Hymes submitted that "the BCT will be employed with multiple BCTs under the C2 [command and control] of a two-star headquarters [division] for medium- to large-scale conventional conflicts. Distributed operations could play a large role in how they are deployed, but it would depend on the situation." He added that, "although it appears non-linear, it really doesn't seem much different then conventional operations today" (Hymes 2006). Likewise, Army Colonel Randolph Binford used the term "amorphous" to describe how he expected the Army to employ future BCTs (Binford 2006). Army Colonel Marcus De Oliviera echoed his peers, explaining that the distributed operations model is indeed "one method of employing BCTs" and agreed that, in accordance with this model, BCTs will be deployed in a non-contiguous and non-linear fashion (De Oliviera 2006). Army Colonel Robert Akam went a bit further, explaining that he expected the BCT "to be totally employed in the context of our new modular design" (Akam 2006). He explained that, "in mid-level conflict, they will rapidly be deployed and conduct independent operations until the initial threat is defeated and the theater becomes more mature. In high-intensity conflict, they will be deployed based on capability and posture in the Army force generation model as part of a 'plug-and-play' two- or three-star or joint headquarters combat capability" (Akam 2006). Finally, Major Jason Mead explained that areas of operation (AOs) "usually have a reason for that BCT to be self-sufficient when they are apart (a city, key terrain, assembly areas in preparation for future maneuvers, etc). Also, BCTs may be larger or smaller, of course, and are not 'one-sized.' Otherwise, BCTs acting separately and independently is an accepted fact of operations" (Mead 2006).

In general, then, most officers knowledgeable on future BCT employment options recognize the likelihood of distributed operations. Accordingly, Deptula and Dahl asserted that "the battlespace of future conflicts will not be characterized by linear means, but rather by nonlinear and noncontiguous areas of operations—leaders in the Services agree on this statement" (Deptula and Dahl 2003, 283). Thus, a basic assumption in this study is that, during medium- to large-scale military operations, the Army will employ a relevant number of BCTs engaged in combat in a non-linear, non-contiguous, geographically separated fashion. Additionally, it is likely that certain concepts discussed in this study would remain somewhat applicable in the event these brigades were to fight in a more traditional manner. Figure 2.2 provides a simplified version of Figure 2.1 that depicts BCTs conducting distributed operations, further amplifying the non-contiguous nature of likely future ground operations. This visualization suggests that the new BCT construct will enable a shift from the linear battlefield of the past to a 360-degree environment in the future.

Fighting in the BCT construct is not simply some distant future vision. The Army has utilized BCTs in Iraq and Afghanistan since U.S. ground forces were employed in large numbers during those operations. In 2006 Testimony before

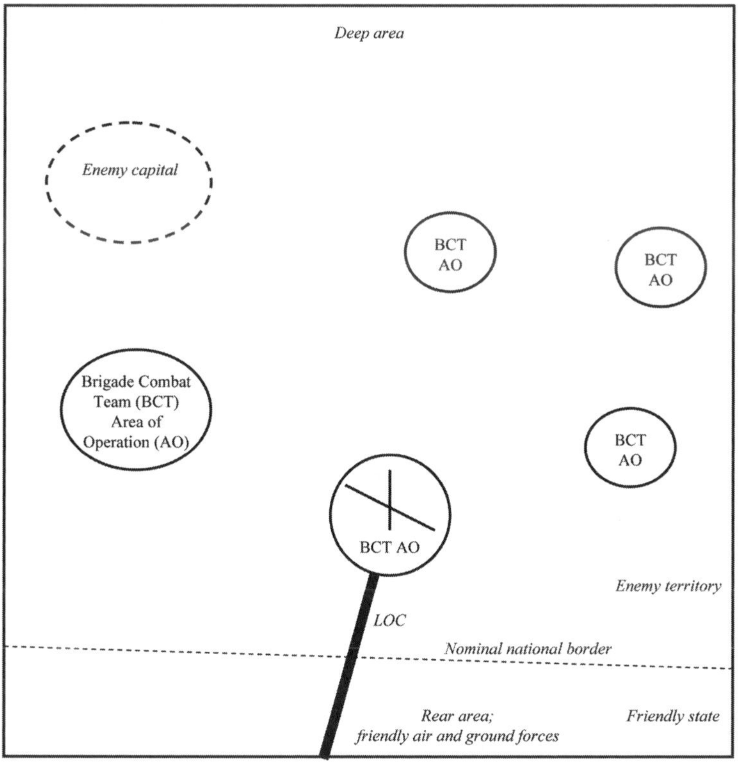

Figure 2.2 Non-contiguous BCT employment

the House Armed Services Committee, Schoomaker cited that, "for the last five years, a period longer than the Second World War, the Army has had as many as 18-20 brigade combat teams deployed on a rotational basis in combat operations" (Schoomaker 2006, 2). The Air Force has been actively engaged throughout the same period in both Operation Enduring Freedom (OEF) and Operation Iraqi Freedom (OIF). It is clearly necessary, then, to develop a sound and coherent body of concepts for how these air and ground forces can fight together. Of course, many ideas have been developed in Iraq for fighting jointly in an irregular warfare environment. However, Schoomaker explained in the same testimony that his Service must "sustain the full range of our current global commitments (which extend well beyond operations in Iraq and Afghanistan)" (Schoomaker 2006, 4). As such, a similar set of concepts must be developed for wars that will be fought in a more conventional manner.

It is also important to note that, during major combat operations in OIF, BCTs fought in accordance with the distributed operations model. For example, the 101st Airborne Division's 1st Brigade, attacked Najaf from the south, while 2nd Brigade conducted the division's longest air assault ever to secure a geographically separated area to the north of the city (Atkinson 2004, 195-197). Williamson Murray and Robert Scales offered a description of the fight for Baghdad suggesting that BCTs of the 3rd Infantry Division (ID) at times fought quite autonomously as well. While they were often in close proximity and even passed through each other on occasion, BCTs were at times geographically separated from the bulk of the American ground forces. First brigade of 3rd ID, for example, conducted an air assault in a series of landing zones southwest of Najaf while elements of 2nd BCT, 3rd ID were to the north, northeast, and west of Baghdad (Murray and Scales 2003, 198-205).

Murray and Scales reported that these BCTs "had relatively little armor" and, as such, resembled Marine ground forces traditionally dependent on the air combat element of the MAGTF (Murray and Scales 2003, 198-205). While the Army's previous division-based organization required support from the air, then, the new construct featuring fewer forces and less armor is far more dependent than before. In essence, *AirLandBattle21* could be viewed as an Army and Air Force version of the more tactically-focused MAGTF, integrating two Services into one fight for a specific purpose in the battlespace. The Marine Corps has always held that its smaller, lighter ground units were inextricably tied to and dependent on indigenous Marine air for success. This, of course, is the cornerstone of the MAGTF concept. Likewise, the Army and Air Force must be welded into a single fighting team to ensure survival and success in the Twenty-first century. An important distinction, however, may be that the Marine Corps, in the absence of large quantities of organic artillery and tanks, has traditionally and purposefully employed its airpower predominantly in the role of support to its light ground forces. *AirLandBattle21* will address the option of support to ground forces but also envisions far more potent opportunities for the integration of airpower and has the specified purpose of providing alternatives to this mindset as the Army lightens its load.

During initial combat operations in Iraq, Murray and Scales wrote that each BCT "was essentially a self-contained close combat unit which, thanks to the speed and killing power of Bradleys and Abrams tanks, had the ability to command as much ground as an entire division during the Cold War" (Murray and Scales 2003, 97). It is interesting to note that, during OIF, brigades were assigned zones analogous in size to areas assigned to entire divisions during the Cold War. Lieutenant Colonel Curtis Anderson, who served in a brigade combat team during initial major combat operations in Iraq, described the zones assigned to individual brigades during Iraqi Freedom: "Each brigade had a zone; it was huge." His brigade, 1st BCT, 1st Armored Division, for example, "had a good, large portion of Baghdad." While brigades technically shared a mutual boundary, individual BCT sectors were, in his words, "humongous." He recalled, for example, that the closest brigade was fifteen miles away and explained that such a zone was "a large space for a brigade to control" (Anderson 2006). This scenario, prevalent in OIF and likely in future conflicts, would clearly lend itself to integration of air and land forces. This transition to brigades conducting distributed operations in division-sized AOs drives the need for and is absolutely central to concepts embodied in *AirLandBattle21*.

Airpower and the BCT

As a result of current and projected autonomous and geographically separated employment, BCTs have become increasingly dependent on airpower. Non-contiguous BCTs will be far smaller in size than their divisional predecessors, will generally have less armor in each unit, will not have the benefit of calling on ground forces from adjacent units when in trouble, and will have a substantially reduced number of artillery pieces at their disposal. Anderson explained that, "as they change our weapon systems to make us lighter, we're not as lethal as we used to be. We get to the fight quicker but we're not as lethal as in the past. So we have some vulnerabilities" (Anderson 2006). The Army transformation, for example, calls for an overall drop of 20 percent in artillery and multiple launch rocket systems and directs a reduction heavy artillery pieces by as much as 60 percent ("Army Transformation and the Impact on the Air Force" 2006, 1). Each of these changes underscores the greatly increased importance of airpower in the battlespace of the future. Hymes acknowledged this new dynamic:

> Transformation has placed the BCT at the center of focus as the basic building block for how the Army fights. In the past the Army was more division centric. Airpower continues to be important for the BCT, but more so in situations where it may be operating independently. Air support will always need to be taken into consideration for a BCT to operate in a distributed manner especially if it is operating apart from the Division (Hymes 2006).

The Army's reduction in field artillery that accompanied its transformation provides an illustrative example of the increased interdependence between air and ground forces. Army Colonel Peter J. Zielinski, a field artillery officer who served as Training and Doctrine Command's (TRADOC) Chief of Concept Development from 2001-2004, explained that, "in the joint fires mission area … the Army can rely upon the contribution of other Services to provide fires when it does not have that capability present in theater at the time when it's needed. That's interdependence. Interdependence allows less redundancy—but not zero redundancy" (Zielinski 2006). For the most part, Zielinski viewed airpower's Twenty-first century contribution as "not necessarily different from the past but certainly of magnified importance in the future. The Army's decision to downsize field artillery was made with the understanding that we will have increased dependence on Air Force capabilities in the area of joint fires" (Zielinski 2006). St-Pierre, who personally briefed former Air Force Chief of Staff, General John Jumper, on the BCT concept, found airpower even more relevant as the Army decreases its organic firepower:

> The Air Force role in the brigade fight is much bigger than it ever was before … it's huge now. Unlike former divisions, the BCT doesn't have the general and reinforcing fires so, in the new design, the brigade commander needs something beyond his indigenous capability to shape the battlefield. In general terms, he would likely devote organic assets such as artillery and Army aviation to the immediate fight and use the Air Force to shape the deeper fight. Now more than ever, the Air Force is truly an integral part of the brigade fight (St-Pierre 2006).

The kingpin of the Army transformation is a rapidly deployable modular force. In its effort to meet drastically reduced deployment timelines, though, the Army will compete with the Air Force more than ever for the same heavy intertheater airlift assets. Additionally, the brigade combat team will be ready to fight at the deployed location in days rather than months. Consequently, the Air Force will likely have less initial airlift available for deployment of its own assets and much less time to position those forces in theater. Furthermore, the result of Air Force transformation will be technologically superior aircraft such as the F-22, but fewer overall combat aircraft will be available for military operations. In order to finance these improved platform capabilities, the Air Force has begun a substantial reduction in personnel as well. It is highly likely, therefore, that the Air Force will have far fewer combat assets in theater at the outset of conflict. The Army has begun to develop concepts for fighting with less equipment and fewer Soldiers. This study attempts to offer ideas on ways both Services can secure success despite reduced numbers of both air and ground forces.

A smaller Air Force, coupled with a more vulnerable Army, suggests that the future employment of airpower will need to be even more efficient and as, if not more, effective than it was in the Twentieth century. This emerging reality requires Airmen to look at old problems in new ways—the essence of a transformation in ideas. Air Force leaders have traditionally viewed their national security roles from the strategic perspective. Air Force doctrine has typically characterized Service missions in a global context, particularly with advent of new technologies that enable impact at this highest level. Interestingly, though, the same airframes and technological capabilities that enable global impact and strategic perspective can now allow the Air Force to make equally substantive contributions closer to the battlefield.

Where you stand, the saying goes, depends on where you sit. Sitting in a cockpit provides a profoundly unique perspective on surface combat operations. This broad viewpoint of war served the Air Force well throughout the last century. However, Twenty-first century warfare requires an integrated approach which bridges the gap between air and ground perspectives. Fortunately, formerly stark differences between a pilot's high altitude vantage point and a brigade commander's view at zero feet can now be reconciled. Remarkable recent advances in sensor technology now allow Airmen to scale their attention all the way down to detailed assessment of a single combatant's weaponry in the field below.

Airmen are capable of viewing the expanse of the battlespace at widely varied levels of fidelity, and this unique capability forms a link in the previously uncharted territory between air and ground. Critically important now are ideas that explore ways to erase the seams between these two areas of responsibility. The development of transformational concepts that integrate air and ground forces is an essential step in bringing this new paradigm to fruition. Metaphorically speaking, the area between the BCT and the aircraft above represents a current intellectual void that *AirLandBattle21* aims to fill. While such efforts may generate crucial thought at the operational level, the impact of these new ideas will rival the strategic airpower theories of the Cold War era.

In place of fewer aircraft and fewer people, then, the Air Force must develop sound concepts that enable success in this environment. Similar budgetary issues in the last century saw periods of competition rather than cooperation between the two Services. Instead of the past parochial fight *against* the Army, the Twenty-first century Air Force would be well advised to focus its efforts on the joint fight *with* the Army. Rather than fighting against their sister Service over dwindling budgets to fund Service-specific transformation, the Air Force and Army should work together on concepts merging these transformed air and ground forces.

One initiative that exemplifies mutually beneficial concept development is the emerging notion of overmatching force. Sir Ian Forbes, retired Royal Navy admiral, explained the theory: "The emphasis shifts away from numbers—and a doctrine of overwhelming force—to take account of harnessing joint capability to decision superiority and thus deliver decisive effects" (Forbes 2005, 146). Overmatching force, then, replaces overwhelming force and, in essence, supplants the traditional definition of *mass* in the principles of war. Shifting from heavy divisions to lighter and more lethal BCTs is a superior example of this concept. The *National Military Strategy* cites the importance of integration, simultaneous operations, and the application of overmatching power. It acknowledges "the concept of surging capabilities from widely dispersed locations to mass effects against an adversary's centers of gravity to achieve objectives" (*National Military Strategy* 2004, 8). Leveraging such ideas will pay great dividends as the Services move forward in this fiscally constrained environment. As the U.S. military moves in this direction, though, it will assume an increased level of risk. As such, a shift away from overwhelming forces on the ground mandates the development of sophisticated concepts to ensure success. It also requires the realization that central to the idea of overmatching force is the mutual employment of air and ground forces in innovative ways.

In a 2001 essay, Michael McMahon warned that, "in developing its transformation strategy, the Army must ensure that it does not focus too heavily on technology itself. It will simultaneously develop the doctrine and training strategies, build organizational frameworks, and educate future leaders to deal with the complexities they will face" (McMahon 2001, 198). In line with this recommendation, the Army has begun to develop some promising ideas for how U.S. ground forces will fight in tomorrow's battlespace. The distributed operations model might serve as an innovative example of an Army concept that rounds out new technologies such as the Stryker vehicle and new organizations like the BCT. The Air Force has developed some similar linkages between technology, organizations, and ideas. There are far fewer coherent, detailed concepts for new Army and Air Force technologies and organizations fighting together.

By looking into the past to inform the future, Williamson Murray found "the problems of integrating new technologies into doctrine and structures that can realistically address the wars of the 21st century will remain as difficult as they were in the last century. But as the Army leadership has grasped, the Army has no choice but to transform" (Murray 2001, 9). With this keen observation, Murray underscored the indelible link between new technological capabilities, new organizational structures, and new doctrinal thought. He also acknowledged that the changed strategic context of the new century mandates transformation in each of these areas. The Army vision speaks of being "ready and relevant." The purpose of *AirLandBattle21* is to ensure both the Army and the Air Force are relevant to Twenty-first century warfare and ready to fight together in the changed strategic environment. Relevance and readiness are not only measured by the operational capability of a Service's new technology and the organizations designed to house them. They are also measured by the transformational ideas that enable success by pulling the technology and organizations of individual Services together into one seamless fighting force.

Chapter 3
Battlespace Transformation

> The dogmas of the quiet past are inadequate for the stormy present. The occasion is piled high with uncertainty, and we must rise to the occasion. As our case is new, so must we think anew, and act anew. We must disenthrall ourselves, and then we shall save our country.
>
> —Abraham Lincoln

At the 2004 Olympic Games in Athens, Greece, the previously dominant U.S. basketball team lost three times in games against Lithuania, Puerto Rico, and Argentina, narrowly clinching a bronze medal in the process. America had previously ruled the court throughout the modern Olympic era, losing only one game in recent history. In Athens, though, independent-minded NBA stars determined to claim victory single-handedly demonstrated repeatedly that even the world's best basketball player cannot win games by himself. Simply stated, these athletes failed to remain preeminent on the court because they did not play as a team. American basketball's foray into Olympic mediocrity sheds a different light on the Defense Department's claim of "U.S. predominance in traditional warfare" (*National Defense Strategy* 2008, 21). It also suggests that bringing home gold in future Olympic games, and remaining unrivaled as the world's supreme conventional military force, depends now more than ever on individual Services and individual athletes emphasizing the value of teamwork. In the past, games could be won by superior individuals simply playing on the same court; wars could be won by individual Services simply participating in the same conflict. As opponents become more adaptive and competitive in the Twenty-first century, playing as a team is increasingly becoming the only sure way to win. This chapter explores concepts designed to allow the Services to do so in conflicts that lie ahead.

The *Air Force Glossary* defines *counterland* as "operations conducted to attain and maintain a desired degree of superiority over surface operations by the destruction, disrupting, delaying, diverting, or other neutralization of enemy forces." It goes on to assert that "the main objectives of counterland operations are to dominate the surface environment and prevent the opponent from doing the same" (*Air Force Glossary*). Air Force doctrine specifies two missions in the counterland category: *close air support* (CAS) and *air interdiction* (AI). The *DoD Dictionary*, which does not recognize the term *counterland*, defines CAS as "air action by fixed- and rotary-wing aircraft against hostile targets that are in close proximity to friendly forces and that require detailed integration of each air mission with the fire and movement of those forces" (*DoD Dictionary*). Joint

Publication 1-02 defines AI as "air operations conducted to divert, disrupt, delay, or destroy the enemy's military potential before it can be brought to bear effectively against friendly forces, or to otherwise achieve objectives." It further specifies that "air interdiction is conducted at such distance from friendly forces that detailed integration of each air mission with the fire and movement of friendly forces is not required" (*DoD Dictionary*). In simple terms, then, air interdiction refers to any attacks against fielded military targets that do not qualify as close air support.

Close Air Support

When Soldiers think about Air Force airpower, they typically envision CAS. In fact, Deptula and Dahl found that "the mindset of 'it's all CAS to me' continues to exist among many ground warriors in the field, leading to disagreements and consternation between Soldiers and Airmen—but much worse, it can have negative consequences in terms of optimally employing our respective forces in joint operations" (Deptula and Dahl 2003, 285). Of note, all fourteen aircraft expending weapons in the Army's *Operating Concept for Operational Maneuver* depicted at Figure 2.1 in Chapter 2 are labeled as "CAS" (*The United States Army's Operating Concept for Operational Maneuver: 2015-2024* 2006, iii).

Close air support was critically important in the Twentieth century and will become even more significant as the transformed Services become increasingly interdependent. In addition to the substantial reduction in field artillery, the shift from divisions to lighter, autonomous BCTs will likely reduce an individual brigade's access to organic fires to an even greater extent. While the full impact of this change is not yet clear, it is likely that this artillery drawdown will increase the Army's requirement for Air Force CAS. According to an Air Force March 2006 *Doctrine Watch*, "due to a decrease in organic artillery firepower and anticipated operations conducted by small units on a more dispersed nonlinear battlefield, the Army has stated a requirement for increased AF CAS support." In fact, wargames designed to conceptualize military operations in 2015 indicated the need for as many as 500 daily CAS sorties in a division-size AO ("Army Transformation and the Impact on the Air Force", 2006, 2).

It is generally safe to argue, then, that the new BCT construct will place an increased requirement on close air support. Furthermore, Army transformation, paired with vastly improved airpower capabilities, suggest that Twentieth century CAS concepts may be ripe for change. In a 2002 Army Command and Staff College thesis, Air Force Major Ralph Hansen found aspects of CAS doctrine written for the Cold War to be incompatible with the transformational technologies and organizations of the new century:

> America's joint forces will operate on this new information battlefield using CAS doctrine that has remained essentially unchanged since the late 1940s. Nonlinear battlefields, digital data links, beyond visual range and GPS-guided

> weapons, and nontraditional air support aircraft have all changed the CAS paradigm. Challenges and questions facing the CAS mission today did not exist even fifteen years ago (Hansen 2002, 7).

A more holistic look at the counterland mission could yield new airpower concepts more harmonious with the non-linear nature of future ground battle.

Air Interdiction and Battlespace Air Operations

Recent events provide a lens with which to view these emerging ideas. Aspects of the 2003 Iraq offensive highlight the deadly synergy between ground and air elements. On 26 March 2003, a column of between 70 and 120 Iraqi tanks, under the expected cover of a heavy sandstorm, broke out of the city of Basra towards the Faw Peninsula. Lured there by the presence of Coalition ground forces, and with the intent to engage them in battle, this convoy of troops and tanks streamed southeast, only to find themselves suddenly targeted by a massive assault from the air. Royal Air Force strike aircraft appeared in the sky above, pounding them with guns and precision munitions. This attack, which continued late into the night, decimated the Iraqi tank formation (Richburg and Glasser 2003, 1). Reminiscent of the "highway of death" on the road from Kuwait City to Basra twelve years earlier during the closing hours of Desert Storm, this operation underscored the overwhelmingly destructive power of air attack against an enemy in the open. Even more importantly, it demonstrated the synergy that occurs when ground presence prompts an enemy to maneuver into a position vulnerable to airpower.

The JFC did not intend for the enemy to maneuver as it did, and Coalition aircraft were not preplanned or "on deck" in preparation for this assault. Instead, the opportunity simply presented itself to the JFACC, and the results were both historic and extraordinary. The most significant lessons to be learned from this experience go beyond an emphasis on airpower's flexibility and destructive effects. The unplanned outcomes provided by this and other examples in modern warfare indicate that campaign planners may now have the unprecedented ability to *intentionally* create situations similar to those surrounding the 2003 Iraqi Basra convoy attack by using a combination of land- and air-maneuver forces to deliberately manipulate and destroy the enemy. Operations since 1990 indicate that maximum efficiency at the operational level of war can be obtained through operational maneuver using land forces to channel an enemy into positions and concentrations most vulnerable to air attack. What is becoming more clear is that the effects created by the joint employment of both air and surface forces together is greater than both could accomplish individually, and that a JFC can expect dramatic results by deliberately using his resources within this construct.

To fully realize the promise of this concept, a mission beyond the traditional ground attack roles of CAS, AI and strategic attack may be necessary. By definition, CAS features aircraft in direct support of ground maneuver. The Twenty-first century may require a transformational mission that allows airpower to serve in the supported role with ground forces initially in support. Similarly, strategic attack favors the use of airpower independent of ground forces. There may be a need for a mission that allows for the *inter*dependent application of air and ground forces. In the Twentieth century, many Soldiers viewed airpower strictly as support to ground forces. Similarly, Airmen often tended to favor the application of airpower independent of the ground fight. These two traditional views of airpower led to a situation where CAS and strategic attack received the bulk of emphasis from a ground attack perspective, leaving a large gap in between.

At first glance, it would appear that the traditional AI mission addressed this void between the close fight and the deep area. On the contrary, the term *interdict* is defined in Webster's with words such as "forbid" and "prohibit." It implies the use of firepower to cut off rather than to destroy, referring more to attacks on enemy lines of supply and communication instead of the enemy forces themselves. The AI mission, then, literally implies attacks that would keep enemy ground forces from reaching the battle. Applying AI by the letter might lead some away from actually attacking those forces. Planners, for example, may select targets that divert, disrupt, and delay—rather than destroy—the enemy. More specifically, preferred targets may be bridges rather than tanks. As such, affected enemy forces might not make the current battle but would potentially remain intact as a credible force to fight another day.

As such, a basic argument laying the foundation for *AirLandBattle21* counterland concepts is that the AI mission as currently defined does not fully address the totality of ground warfare in the Twenty-first century. In a functional sense, it falls short of fully embracing the enemy's fielded forces as a primary target. Geographically, air interdiction too broadly encompasses a remarkably wide area that ranges from just beyond the close battle to the far reaches of the entire battlespace, even incorporating applicable targets in populated areas and the adversary nation's capital. By specifically acknowledging the existence of *strategic attack* and clearly defining the term, the Air Force has addressed one end of this spectrum. The other end warrants similar attention. Of note, there is no joint definition for strategic attack. Within the joint community, therefore, air interdiction even more broadly encompasses everything beyond CAS.

Consequently, absent and necessary in the new millennium is a well-defined doctrinal counterland mission, and operational concepts that accompany it, for attacking the enemy's fielded forces before they close for direct battle with friendly forces. There is a need for a mission that specifically recognizes airpower's capacity to not simply interdict but to kill these forces in large quantities. Beyond independent-oriented strategic attack and support-oriented CAS, the requirement exists for an airpower mission that allows air and ground forces to fight *inter*dependently with airpower at times in a supported role.

AirLand Battle of the Cold War era featured a vital airpower mission called *battlefield air interdiction* (BAI). This mission was central AirLand Battle's basic tenet of attacking second-echelon Soviet follow-on forces in the Fulda Gap before they closed with friendly forces. As the Berlin Wall came down, AirLand Battle fell away and this mission disappeared. In an effort to address the resulting gap, *AirLandBattle21* advocates the advent of a new airpower mission labeled *battlespace air operations* (BAO) that will enable closer integration of Twenty-first century air and ground forces. While the idea has been discussed in certain airpower circles, it has not yet gained traction. Following OEF, Deptula strongly supported a concept very similar to BAI. His transformational variant capitalized on dramatic advances in sensor capability, command and control, and precision engagement, and applied powerful air assets against enemy fielded forces on the battlefield to achieve success without the need to prematurely expose friendly ground forces to the destructive effects of the enemy. Air Force Major Terrance McCaffrey further codified this movement in a 2002 School of Advanced Airpower Studies thesis, concluding that "I believe that there is a doctrinal void that needs to be filled and my proposition for filling it is battlefield air operations" (McCaffrey 2002, 102). Because the term *battlespace* is now widely accepted across the Services and the joint community, this study uses the term *battlespace air operations* in place of *battlefield air operations* used previously.

This emerging concept of BAO could be viewed as the product of a long evolution of airpower thought and experience. Between the closing hours of Desert Storm, and the opening shock of Iraqi Freedom, shrinking defense budgets and legitimate differences of opinion led land-centric and air-centric warfighters down a contentious path, with strongly-held viewpoints institutionally polarizing the debate between both camps (Lambeth 2000, 296). At the extremes, advocates on each side fueled a well-intentioned squabble over the best ways to enhance the nation's war-winning capability. Ground-oriented advocates decried the limitations of airpower, echoing a "must-have-boots-on-the-ground" mantra with an almost defensive zeal, while airpower zealots occasionally oversold the benefits of strategic air attack, sometimes overlooking the validity of fielded forces as a strategic target. As McCaffrey observed, "after Desert Storm, what Joint Force commanders were missing was a coordinated synergy of air and ground combat effects that would enable them to create the most overwhelming dilemma for an enemy" (McCaffrey 2002, 116). During this period, Coalition operations verified the unique ability of friendly surface forces to compel enemy reaction, and at the same time, highlighted the maturing ability of airpower to apply precise, destructive effects against an enemy.

The term *battlefield air interdiction*—the Twentieth century predecessor to BAO—originated in NATO in 1979, and developed doctrinally for both the Army and the Air Force through 1990. Both Services' doctrine demonstrated a common understanding of the BAI mission during the 1980s (McCaffrey 2002, 3). Prior to Desert Storm, the Army priorities focused on targets beyond the front lines whose destruction supported and facilitated the friendly ground scheme of maneuver. Similarly, BAI was significant to the Air Force because it allowed aircrews to engage targets between an established fire support coordination line (FSCL) and the line of friendly forces, without distributing control of small packets of airpower to lower-level ground commanders (McCaffrey 2002, v). Because of this shared area of interest, the BAI construct created a battlefield zone which allowed for mutual coordination and cooperation and, in theory, linked the two Services in the targeting of enemy forces located in the shallow region between friendly ground forces and the FSCL (McCaffrey 2002, 3). BAI differed significantly from the CAS mission which required detailed, real-time integration with ground controllers due to the immediate proximity of friendly troops to the intended targets.

Because AirLand Battle and its associated BAI concept had gone the way of the Cold War, the BAI mission was never included on the air tasking order (ATO) during the first Gulf War. An era of doctrinal friction between the Army and the Air Force over the areas BAI addressed ensued throughout the remainder of the 1990s (Palmer et al. 1998, 3). The *Gulf War Air Power Study* (*GWAPS*) found that the BAI would have been compatible with the Desert Storm air campaign. However, the air component chose to identify "missions inside the fire support coordination line [as] CAS missions and all others outside the line, Air Interdiction (AI) missions, which deleted BAI as a type of mission" (McCaffrey 2002, 46). As the ground campaign drew closer, this change made it more difficult for ground commanders to include their priorities on the ATO and established the FSCL as a barrier between land and air control rather than a measure to enhance coordination and create synergy (McCaffrey 2002, 48).

During Desert Storm, the JFACC had airpower to spare. Because ground forces rarely engaged in close contact with the enemy, there existed an excess number of available and unassigned CAS sorties. Unfortunately, elimination of BAI from the table prevented the Coalition from engaging some Iraqi formations efficiently with these unused aircraft. To achieve their own objectives, ground commanders sought to push the FSCL as far forward as possible in order to enhance their own ability to affect the extended battlespace, which unintentionally restricted aircraft designated for CAS missions from engaging the enemy. This arrangement required extensive ground coordination shy of the FSCL but, because friendly forces were not in contact with the enemy, ground controllers had no CAS targets for these aircraft. As GWAPS reported, "it is clear that there was such an unintended zone in which Iraqi forces benefited from the shortcomings of Army/Air Force coordination" (McCaffrey 2002, 59). Each Service, struggling to optimize its own contribution to the joint fight in the shallow zone forward of friendly troops actually mitigated the combat effects of both Services, resulting in missed opportunities in the battlespace. Despite the shortcomings created by different doctrinal viewpoints, this fundamental gap in the U.S. joint warfighting construct persisted through Operation Enduring Freedom, and into the days immediately prior to the second Gulf War.

In the absence of a construct like BAO, many workarounds emerged to address shortfalls in the AI mission area. One such ad hoc substitute has been this growth of the CAS mission well outside its formally defined intent. In an interview with McCaffrey, for example, Deptula explained the use of airpower during OEF in the context of BAO: "I think we're doing battlefield air operations, we're just calling it something else … right now we're calling it CAS, and it's not CAS" (McCaffrey 2002, 102). Deptula and Dahl further explained that, "in the Afghan 2001 and Iraq 2003 campaigns, there were many scenarios in which we employed airpower as an element of those joint campaigns to achieve the JFC's goals that did not involve troops in contact. Iraqi Freedom also saw massive use of SOF forces in a more conventional role. These scenarios don't fall within the definitions of CAS and AI" (Deptula and Dahl 2003, 24). The concept of BAO, then, fills both a theoretical gap in current doctrine as well as a practical gap in the battlespace. Deptula and Dahl concluded that "establishing BAO as a separate mission will bridge the gap between the traditional, linear battlespace of the past and the reality of the nonlinear, noncontiguous and nontraditional battlespace of today and tomorrow" (Deptula and Dahl 2003, 25). By providing planners with a coherent mission beyond CAS, AI, and strategic attack, BAO fills the current void in both theory and practice.

Attempting to clearly define BAO while exploring this idea in 2002, Colonel Gary Crowder led the development of a proposal that reflected the new construct. According to Crowder, BAO was defined as "Air Operations conducted in a supported role against an enemy land force with the intent to destroy or neutralize them" (Crowder 2003). This proposal sought to fill a doctrinal gap between the traditional AI and CAS missions, and was intended to describe air attack against surface forces in those situations when friendly troops are absent, or where their proximity to adversary ground forces would not require detailed integration inherent in the CAS mission.

Beyond the Fire Support Coordination Line

One pragmatic approach used largely as a result of this past void has been the practice of assigning aircraft to attack targets of opportunity in a particular grid on the ground, called a *kill box*. Accordingly, some may argue that the kill box has replaced the BAI mission. More accurately, though, the kill box concept is, at its core, a fire control measure without a doctrinal mission and without the associated widely accepted operational concepts to perform that mission. It would be analogous to simply recognizing the existence of a FSCL without identifying CAS as the associated mission and without developing the robust joint doctrine, employment concepts, and the many detailed tactics, techniques, and procedures (TTPs) that accompany the CAS mission. The Air Land Sea Application Center (ALSA), an inter-Service organization that provides multi-Service TTPs which serve as an inventive workaround to inherent challenges in the joint doctrine process, studied the kill box concept. In the preface to their 2005 publication on kill box employment, ALSA cited that "there is no formal kill box doctrine or tactics, techniques, and procedures" and explained that their multi-Service document "updates the definition and establishes the kill box as a fire support coordination measure (FSCM)" (*Kill Box: Multi-Service Tactics, Techniques, and Procedures*

(TTP) for Kill Box Employment 2005, i). The kill box manual observed that, on a more traditional linear battlefield, "kill boxes can augment use of traditional FSCMs, such as fire support coordination lines (FSCLs) …" The ALSA document went on to explain that, in future non-linear battlespace, "when traditional FSCMs are not useful or are less applicable, the kill box can be another method for identifying areas to focus air and indirect fire assets" (*Kill Box TTP*, I-4). It specified that "the kill box may be an applicable tool where traditional coordination measures (e.g., FSCL) do not exist or have not been established" (*Kill Box TTP*, II-1). Prior to the publication of this document, the joint community defined a kill box as a "three-dimensional area reference that enables timely, effective coordination and control and facilitates rapid attacks" (*DoD Dictionary*). As promised, the ALSA study offered a refined multi-Service definition: "A kill box is a three-dimensional fire support coordinating measure (FSCM) used to facilitate the expeditious air-to-surface lethal attack of targets, which may be augmented by or integrated with surface-to-surface indirect fires" (*Kill Box TTP*, I-1). This analysis went on to clearly indicate that the kill box concept is an FSCM without a body of concepts or doctrinal procedures and without a codified and clearly defined mission.

As another workaround to the absence of a mission such as BAO, planners in Korea developed an ad hoc remedy they called *mobile interdiction* (XINT). In the absence of an applicable mission, the term XINT was typically used to describe the task performed in the kill box as assigned by the ATO. Consequently, where the kill box constituted an interim FSCM, XINT might be viewed as an improvised mission. While XINT filled the void in that theater for some time, it is not used widely elsewhere and does not appear in either Air Force or joint doctrine. Furthermore, while resembling a mission in the traditional sense of the word, it is not accompanied by widely-recognized and well-developed operational concepts.

During OIF, planners arrived at a similar ad hoc concept by dividing Iraq into a grid of kill boxes and directing swarms of airpower against those containing Republican Guard units. In name, this mission was identified as "KI/CAS, for 'kill box interdiction close air support'" (Scarborough 2003, 5). It could have more accurately been labeled BAO. Instead of XINT and KI/CAS, however, the 2006 Air Force counterland publication established *XAI* as "a mission that pursues a designated area versus a particular target", explaining that "XAI may fly airborne alert or search particular areas to strike at targets of opportunity" (*Counterland Operations* 2006, 7). In establishing this term, Service doctrine seemed to step in a positive direction by recognizing the lack of a current mission to accompany the kill box FSCM construct. However, this initial movement fell short of clearly delineating a new mission between CAS and AI and did not offer the necessary body of detailed procedures. Furthermore, the acronym XAI carried with it the previously-discussed baggage of the word "interdiction" as literally defined and commonly practiced. Finally, the manual did not fully recognize the critical need for an approach that would allow for the full integration of Twenty-first century air and ground forces.

Largely as an outgrowth of this mission gap, the original direct doctrinal tie between CAS and the FSCL is arguably no longer fully intact. This development further complicates an already convoluted coordination system. For example, while counterintuitive, current employment methods allow for air interdiction missions to be flown in areas short of the FSCL. A driving factor for this practice is the fact that ground commanders have typically established the FSCL well beyond the effective range of their organic artillery assets. When the FSCL is drawn unnecessarily distant from friendly forces and the practical range of their indigenous firepower, it requires close coordination designed for different situations and, therefore, often beyond the capability of ground forward air controllers. Because of the greatly increased distances, these controllers often cannot see the target areas they are required to clear the aircrews to attack and sometimes cannot reasonably establish where the aircraft may be aiming their weapons. For example, a 1999 test by the Joint Close Air Support (JCAS) Joint Test and Evaluation (JT&E) found that ground controllers met all necessary criteria under such conditions only 38 percent of the time (Hansen 2002, 34). This situation often results in a sanctuary for enemy forces operating in the area between the FSCL and the effective range of Army organic fires. The 2006 Air Force *Counterland Operations* doctrine document clearly articulated this phenomenon:

> The optimum placement of the FSCL varies with specific battlefield circumstances, but typically it should be placed where the preponderance of effects on the battlefield shifts from the ground component to the air component. In this way, the FSCL placement maximizes the overall effectiveness of the joint force, and each component will suffer only a small reduction in efficiency. To place the FSCL so deep or shallow that one component is given complete freedom to operate usually results in the other components being so restricted that overall joint effectiveness suffers. The proper location for the FSCL may also shift from one phase of the war to the next, depending on the scale and scope of each component's contribution during that phase. FSCL placement must also take into account the ground scheme of maneuver and should be based on anticipated, not current, ground force positions at the time the FSCL will be active. History has shown that placing the FSCL too deep is detrimental to overall joint force effectiveness and may even provide the enemy a sanctuary from effective air attack (*Counterland Operations* 2006, 69).

The placement of the FSCL during Desert Storm provides a fitting historical example. In the closing hours of the first Iraq War, the FSCL was established well beyond the presence of friendly forces and outside the Army's ability to range indigenous fires or coordinate CAS in any effective way. As a result, artificial limitations placed on the Coalition's ability to mass airpower against Saddam's Republican Guard gave sanctuary to retreating forces and unwittingly allowed many forces to survive. Much of the same equipment was used against the Iraqi people after Desert Storm and challenged Coalition forces once again twelve years later during Iraqi Freedom.

Ground commanders have traditionally exhibited an unwillingness to establish FSCLs in accordance with the doctrinal Air Force recommendation. In an effort to maximize flexibility on the ground, they have unwittingly greatly reduced flexibility and efficacy in the air. In their reluctance to assume marginal risk of friendly fire from the air, ground commanders have arguably assumed much greater risk from enemy forces allowed to survive. The likely result in such a scenario would be fewer enemy tanks killed and a greater number of friendly casualties. Despite these high stakes, this outdated remnant of the Cold War has yet to be resolved. In fact, Air Force doctrine identified the problem as recently as Iraqi Freedom:

> OIF employed traditional FSCMs. Because of the Army's extensive process required for changing linear FSCMs, moving the FSCL proved to be a time-consuming process. Therefore, the initial FSCL was placed well beyond the range of land fires in order to accommodate the anticipated rapid movement of land forces into Iraq. The deep placement of the FSCL hampered the efficiency of airpower. Ground forces, and their associated TACPs [tactical air control parties], were incapable of detailed integration beyond the range of their organic fires because no one was able to observe adversary targets. Aircrews were still required to comply with coordination procedures short of the FSCL. The time-consuming clearance process hindered the expeditious attack of fleeting targets beyond the range of the organic artillery. As a result, the area between the maximum range of land fires and the established FSCL created a sanctuary for enemy forces (*Counterland Operations* 2006, 69).

In an effort to address this issue, 2006 Air Force counterland doctrine observed that "the preponderance of kinetic effects shifts from landpower to airpower near the maximum range of organic field artillery." It went on to recommend the placement of the FSCL relative to this range: "Therefore, under all but the most rapid ground maneuvers, the FSCL is normally placed near the maximum range of tube artillery because air and space power provides the most expeditious attack of surface targets beyond that point." Furthermore, "to facilitate a rapidly moving battlefield, a common practice is to establish 'on-call FSCLs' in advance that can be activated as the ground force moves" (*Counterland Operations* 2006, 70). This, however, may be a legacy solution to a traditional problem. Under the new Army construct, an autonomous, geographically separated BCT will have access to a greatly reduced number of artillery assets. Meanwhile, modern airpower has become far more capable of massing effects. As such, discussing the long-range preponderance of kinetic effects may be less applicable to the Twenty-first century BCT fight as it was when large Army divisions held hands in the Fulda Gap. This emerging reality, combined with the inherent linear nature of the FSCL, brings even this more recent remedy into question. To be fair, the same doctrine document went on to recognize the need for an FSCM more compatible with non-linear ground operations. Unfortunately, the Air Force remained wedded as late as 2008 to the kill box as the only practical or doctrinal remedy (*Counterland Operations* 2006, 68, 72).[1]

During OIF, the Marine Corps developed a supplemental FSCM they labeled the *battlefield coordination line* (BCL) (Grant 2004). Typically tied to the range of their artillery, the BCL did in fact delineate CAS from air interdiction, mitigating confusion, unnecessary layers of coordination, and potential enemy sanctuary. A new construct based on this Marine idea would likely remedy what Douglas Johnson, a professor at the Army War College and retired Army colonel, identified as the greatest challenge of the new BCT construct: "The hairiest change is the idea that the Army will actually be able to field a force that, once married to adequate lift, will be able to go leaping about like a scalded frog" (Johnson 2006). This greatly increased ability for autonomous friendly ground forces to freely move across the breadth and depth of the battlespace is wholly incompatible with the linear Soviet-era FSCL. In order to fully merge the useful BCL construct with the proposed non-linear nature of future BCT distributed operations, using *area* rather than the word *line* would be more appropriate. Additionally, as previously discussed, this study uses the joint term *battlespace* in place of *battlefield*. As such, *AirLandBattle21* proposes use of the term *battlespace coordination area* (BCA). This new BCA is not unlike the current doctrinal *joint special operations area* (JSOA), which has been quite successfully applied in conjunction with non-linear special operations. Of note, this new FSCM is fully compatible with the BAO construct as well. On the one hand, it will allow aircrews to perform doctrinally-correct CAS within an appropriately established BCA. On the other, it will enable conduct of the BAO mission when reduced levels of coordination would be more appropriate.

In a 2002 Army Command and Staff College thesis involving CAS and the interim brigade combat team (IBCT), Air Force Major Ralph Hansen drew conclusions pertinent to interdependent warfare in the Twenty-first century:

> The lighter and more agile Army will deploy with far less organic firepower than before. The IBCT will have mobile artillery and mortars, but these lighter

1 The Air Force doctrine document found that, "historically, linear operations have used linear FSCMs such as the FSCL. However, as operations move towards being non-linear, dispersed component AOs necessitate the need for nonlinear FSCMs such as kill boxes. Advancements in data link technology and digital information have increased the potential for combat forces to effectively coordinate and conduct nonlinear operations" (*Counterland Operations* 2006, 68). It went on to submit that, "in conflicts characterized by nonlinear operations, ground forces occupy pockets that may have large distances of open terrain between them (often occupied by the enemy). Under such circumstances, the classic linear concepts such as the FSCL may need adjustment. One option is to create an alternate FSCM based on a common reference system (such as a standardized box, circle, or other easily employed shape) to accomplish the same task that the FSCL performs for the linear battlespace" (*Counterland Operations* 2006, 72). Both citations clearly opt for the kill box concept as a viable solution.

> units have necessitated a reduction in both the number and caliber of fire support assets. This reduction in internal fires capability will force the lighter brigades to rely more on joint firepower on the battlefield of the future, such as fixed-wing CAS and naval gunfire, much as the U.S. Marine Corps does today (Hansen 2002, 8).

Consequently, a concept such as the BCA may be as appropriate for future Army and Air Force air-land integration as the BCL is for the Marine Corps' MAGTF today.

A traditional view of the battlefield depicted in Figure 3.1 features an FSCL that may be aligned with territorial zones of operation but well beyond the areas where associated BCTs actually plan to operate in the near future. Such an FSCL would likely extend past the range of indigenous BCT firepower. Furthermore, applying the same traditional mindset used during Desert Storm and Iraqi Freedom, CAS would span the entire area within, around, and between the BCT AOs. The bulk of this battlespace would consequently feature no hostile action between

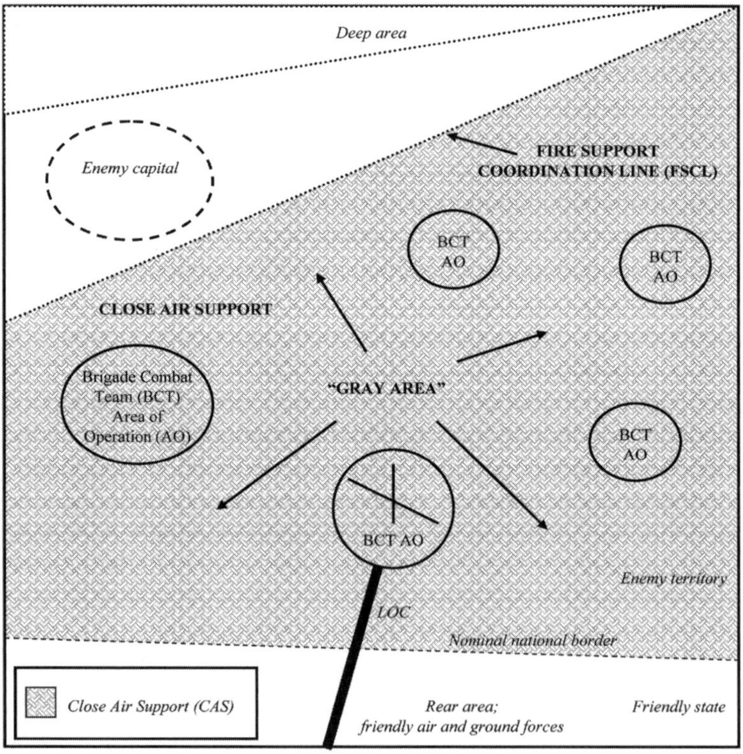

Figure 3.1 A traditional view of the battlefield

enemy and friendly ground troops in close contact. Nonetheless, as in the two Iraq wars, extensive positive control measures would presumably be required in order to avoid friendly fire incidents from the air despite the absence of friendly surface forces in much of the "gray area" between BCT AOs. Consequently, the time-consuming actions required by doctrine, such as the detailed nine-line CAS briefing and positive clearance to expend weapons, would superimpose substantial inefficiencies unnecessary in such a situation. In practical terms, such extraneous procedures would arguably drastically reduce the number of military targets such as enemy tanks that friendly combat aircraft could destroy given time and fuel constraints. Simply stated, the gray area would become a further sanctuary for the adversary's fielded forces.[2]

A transformational view of the battlespace such as the one depicted in Figure 3.2 would also feature areas of close air support. These CAS areas, however, would be carefully planned to encompass only those regions where troops in contact could be reasonably expected and where positive control measures would therefore be necessary and appropriate. In contrast, the BAO areas would require far less restrictive procedural controls similar to kill box methods currently employed. In such a construct, the FSCL from the days of the linear battlefield would be replaced with BCAs far more compatible with the non-linear battlespace of the Twenty-first century. Rather than sanctuaries, gray areas would become killing fields. This transformational concept would allow air forces far greater flexibility in the potentially large regions between BCT AOs where no friendly ground action is planned or anticipated. It would replace unnecessary positive control with appropriate levels of procedural control commensurate with conditions on the ground and desired levels of risk. Instead of CAS and the baggage it brings to the battlefield, the transformational view would allow air forces in these gray areas to perform the far more appropriate mission of BAO. Simply stated, these actions would greatly increase the opportunity for mass destruction of military targets from the air and enable a wide variety of airpower missions for aircraft operating in these regions. This new view of the battlespace also features the traditional airpower missions of air interdiction and strategic attack, nominally positioned in areas where they may be conducted as directed by the combatant commander.

This new paradigm may strike some as radical from the Army's perspective. However, Colonel Pete Curry, Army aviator and Marine Corps War College professor, fully recognized the potential applicability and utility of such a transformational approach in the complex battlespace of the future: "Somebody's going to 'own' that terrain ... the trouble is that a brigade would normally have such a large area in a combat situation that there is a lot of 'gray area' in the guy's own zone." Curry recognized the feasibility of a ground commander acknowledging "I can't control this whole zone of mine—I don't have enough guys" (Curry 2006).

2 The FSCL does not necessarily always serve as a demarcation between CAS and other airpower missions. The traditional view, however, has generally held that CAS is the predominant aerial mission flown short of the FSCL.

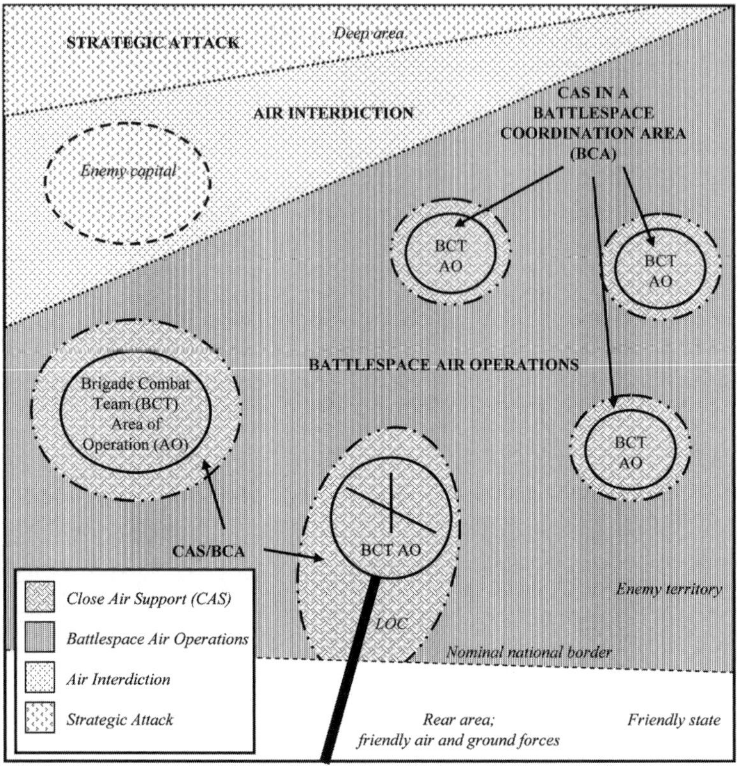

Figure 3.2 A transformational view of the battlespace

A Twenty-first century combat environment, replete with gray areas and short of ground troops, would be just the scenario where the Joint Force Commander might assign certain areas to the air component and direct the ground component to remain clear. The BAO mission and the BCA fire support coordination measure would fully enable such an integrated approach.

Furthermore, as Figure 3.3 illustrates, a grid system could be used in conjunction with the BAO concept. Such a practice would enable air forces to perform CAS in all sectors where it is required without attempting to superimpose the linear FSCL onto non-linear battlespace. Activating BCAs in sectors or map grids would be simple, would allow the Army and Air Force to communicate in a common language, and would enable relatively quick activation and deactivation of these areas. One could view BAO as the mission currently lacking in the concept of kill box employment. Kill Box 2H in Figure 3.3 would serve as a kill box in the traditional sense where BAO would be the mission conducted in this sector. Kill Box 3H draws attention to the possibility that a BCA could dissect a kill box, allowing CAS within the BCA and BAO, with its reduced coordination requirements, beyond the BCA. As mentioned, the concept of a BCA is quite

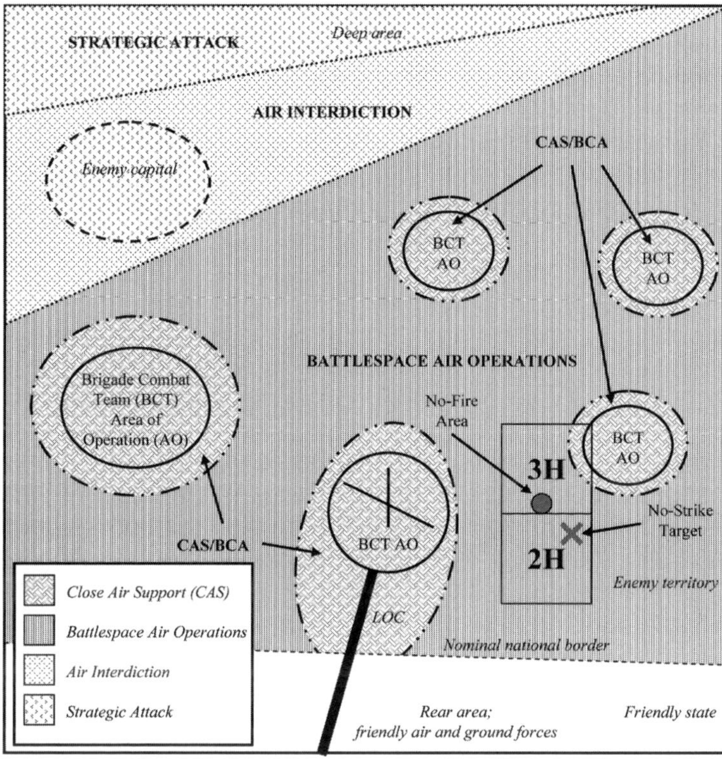

Figure 3.3 Battlespace air operations and kill box employment

similar to a JSOA, which is doctrinally mature and has been used to great effect in many military operations. Consequently, doctrinal and practical transition to the BCA is quite feasible and could be seamless as well. While codifying the change in joint doctrine will be a challenge, the trail blazed by the JSOA concept suggests the likelihood of a successful outcome.

As previously discussed, then, current doctrine lacks a mission, related FSCMs, and associated operational concepts that are fully compatible with the non-linear battlespace awaiting the U.S. military in the Twenty-first century. Using concepts from *AirLandBattle21*, the mission void would be filled by the incorporation of BAO. In the area of fire support coordination, the FSCL of old and the current kill box concept of today would give way to the BCA of the future. The lack of well developed joint doctrine, transformational ideas, and widely accepted inter-Service operational concepts for integrating airpower and BCTs would be met in the next chapter with the concept of air as a maneuver force.

Chapter 4
The Concept of Air as a Maneuver Force

> Those skilled at making the enemy move do so by creating a situation to which he must conform; they entice him with something he is certain to take, and with lures of ostensible profit they await him in strength.
>
> —Sun Tzu

The *DoD Dictionary* offers two definitions for the term *maneuver*. The most general and widely known is the "employment of forces in the operational area through movement in combination with fires to achieve a position of advantage in respect to the enemy in order to accomplish the mission." However, another DoD-approved definition explains that maneuver is "a movement to place ships, aircraft, or land forces in a position of advantage over the enemy" (DoD Dictionary). Of note, there is no mention of taking or holding ground in either definition, although these requirements are often levied on the concept of maneuver when the term is associated with surface forces. Curry alluded to this traditional perspective when he explained his view of maneuver as "an element that will gain positional advantage on the enemy … Some people say you can't gain leverage without a ground force to leverage off of. You can't take a crowbar and hold it in thin air and try to do something" (Curry 2006). Nonetheless, the joint definition directly specifies that aircraft can serve as a maneuver force and sets no requirements that an air-maneuver force (AMF) would be unable to achieve. In fact, the concept of air maneuver has begun to emerge in both Army and Air Force doctrine (See *The United States Army Concept for Tactical Maneuver for the Future Modular Force: 2015-2024* 2006, 21; *The United States Army Operating Concept for Operational Maneuver: 2015-2024* 2006, 25). Less prevalent, however, are novel ideas for how an AMF might be integrated with ground forces and how the new BCT construct mandates transformational ideas that enable interdependent air and ground forces to succeed in tomorrow's complex, non-linear battlespace.

Today, in the aftermath of the major combat operations portion of OIF, the BAO/BCA proposal offers unprecedented options to the JFC. As valuable as BAO can be in the execution of missions *without* the presence of friendly ground forces, its greatest promise may actually lie in the way it can facilitate joint warfare planning to maximize the destructive effects of both ground and air power and serve as the catalyst for a new form of joint synergistic campaign planning. Using BAO, the JFC can intentionally manipulate the enemy with a friendly ground presence, and then apply an air-maneuver force to achieve remarkable effects on the battlefield. Rather than using air in the traditional construct of supporting boots on the ground, the JFC can use those boots to prime the battlefield, shaping the enemy order of battle into a configuration vulnerable to air attack. Used in this way, BAO could serve as the linchpin of a truly joint future campaign. In the case of an enemy surface force, the JFC's desired operational effects will likely be to render that enemy combat-ineffective in the quickest, most efficient way. In the pursuit of this goal, the JFC can direct integrated teams of air and surface forces in dynamic ways to impose an operational dilemma upon the enemy: either he can remain in place and become vulnerable to a joint offensive, or move, exposing his forces to air attack (Bingham 2002, 53).

The combat planner's key to a successful strategy will be to force a reactive posture that affords the JFC his greatest asymmetric advantage. This kind of joint warfighting will exploit joint maneuver, human factors, and achieve a unique synergistic effect by employing air and ground forces in concert to produce effects disproportionate to the expended effort (Bingham 2002, 54). The Air Force's 2006 counterland doctrine publication emphasized airpower's ability to either act in support of friendly surface operations, or as the main effort against enemy fielded forces to achieve these effects (*Counterland Operations* 2006, 3-4). This publication stressed the importance of using friendly surface maneuver in concert with air operations to make the enemy more vulnerable to joint forces (McCaffrey 2002, 94). From a surface commander's point of view, the most significant initial recognition of this construct appeared in the 2001 *Field Manual (FM) 3-0, Operations*: "Army forces may be the supporting force during certain phases of the campaign and become the supported force in other phases" (*FM 3-0* 2001, 2-7). The significance of these two documents underscores recognition within the Army and the Air Force of the need to coordinate actions across the battlespace to maximize destructive effects against enemy fielded forces.

When the JFC directs ground forces either to push or to entice enemy movement, he uses that presence to expose the reacting enemy. As the enemy force masses or moves in an attempt to create advantages against friendly surface forces, it presents vulnerable targets for precision engagement from the air (Bingham 2002, 56). The result is that the land-force commander can intentionally delay active surface engagement until enemy fielded forces are dramatically weakened by the air-maneuver force. The JFC can then direct surface forces to attack at the right time and place to capitalize on this effect. In a somewhat similar approach during OIF Phase III operations, land commanders delayed forward movement between 29 March and 3 April 2003, as Coalition airpower pounded elements of the Republican Guard defending Baghdad and slashed enemy force structure in a veritable rout of fielded forces prior to friendly surface engagement. According to a 2 April 2003 *U.S. News and World Report* article, "like the Baghdad Division, the Medina Division had been judged 'combat ineffective' by the Pentagon—meaning that airstrikes had reduced the unit's combat power by at least 50 percent" (Mazzetti 2003, 1). When the land component moved forward, advancing U.S. troops continued the destruction of Republican Guard forces.

To maximize these effects, the JFC must optimize his capability to use surface forces in conjunction with an air-maneuver force to achieve synergistic gains.

Accordingly, Martin Van Creveld defines seven central elements that help to define the art of maneuver: Tempo, surprise, flexibility, decentralized execution, mission type orders, combined arms, and *schwerpunkt*, or the ability to hit pivotal, vulnerable targets, all represent fundamental essentials of traditional maneuver warfare (Van Creveld 1994, 7). According to Van Creveld, "maneuver seeks to avoid both the battlefield and the bloodshed by moving to the next highest operational level; in other words, it seeks to decide the tactical engagement by using grand tactics, the grand tactical engagement by using operational art, and the operational-sized engagement by resorting to strategy" (Van Creveld 1994, 9). Before a fight takes place, a maneuver force seeks to place the enemy at a disadvantage by moving into a position that maximizes friendly strengths, in an attempt to achieve ultimate victory while engaging in the minimum amount of combat (Van Creveld 1994, 1, 209). Traditionally, maneuver warfare has been a limited, two-dimensional paradigm with an emphasis on the ground scheme of maneuver. Army's *FM 3-0*, however, described maneuver warfare in a way that opens new possibilities for joint operations: "Maneuver implies more than the use of fire and movement to secure an objective; it aims at the complete overthrow of the enemy's operational design. It requires audacious concepts and ruthless execution" (*FM 3-0* 2001, 6-13). Maneuver is no longer just about seizing ground; it is about achieving desired joint effects through bold, inventive plans (*FM 3-0* 2001, 7-6).

In a joint operation against a fielded enemy, airpower can actually perform as a maneuver force in line with classic Army definitions and current joint definitions, applying massed and concentrated firepower against exposed forces. Organizational constructs such as the air operations centers paired with transformational technology recently developed and fielded, allow the air component to mass and concentrate overwhelming airpower at a specific time and place against an enemy's surface forces. Doctrinally, Airmen have long advocated the capability to employ rapid maneuver from the air against critical fielded forces, but advances in technology and organizational force structure have converged in the Twenty-first century to make this once-abstract theory a realistic expectation (*Air Warfare* 2000, 4). In future joint warfare at the operational level of war, the functions and effects both land and air forces can achieve against a fielded enemy will remain quite comparable, due to their similarly destructive capabilities (Givens 2002, 84). Both components have limitations, however, which strengthen the case for integrated air and ground operations. Just as aircraft cannot occupy territory, surface units do not possess airpower's inherent capacity to occupy the battlespace over the enemy (Givens 2002, 85). Although an AMF has the limitation of not being able to hold ground, its unique third-dimensional targeting perspective affords an advantage that forms the essence of maneuver warfare. As General Alfred M. Gray, former Marine Corps Commandant instructed, "warfare by maneuver … stems from a desire to circumvent a problem and attack it from a position of advantage rather than meet it straight on" (*Warfighting* 1997, 37).

This AMF concept opens a new way of viewing joint warfare. Not only can ground maneuver forces reposition to gain relative advantages over the enemy, but an air-maneuver force can be planned, marshaled, and concentrated on the battlefield, in concert with manipulative ground forces, at the precise time and place to achieve overwhelming success. Efficiency may be the most substantial benefit of utilizing an air-maneuver force. After gaining the necessary level of control in the air, a JFC will place fewer American lives at risk and yet achieve the same objective in the field by applying an air-maneuver force to an exposed enemy rather than initially engaging with troops or armor. Using the full extent of their battlespace, joint planners are thus able to exploit all of the available maneuver room to full advantage.

While current Service doctrine in both the Army and the Air Force has begun to embrace the AMF concept, it sometimes falls short of fully grasping the full potential and widespread applicability of aerial maneuver. An October 2006 forward-looking Army doctrine pamphlet, for example, called for "a new tactical paradigm" it dubbed "decisive maneuver." The manual explained the new paradigm:

> Instead of the former pattern of developing the tactical situation in contact, followed by maneuver, often under fire and/or observation, to enable decisive engagement, the future Modular Force will: (1) Develop the situation largely out of contact; (2) maneuver out of contact to positions of advantage, while retaining freedom of action; (3) isolate, immobilize and disorganize enemy forces, denying them the ability to coordinate or cooperate effectively; and (4) conduct decisive combat at the time and place of the commander's choosing (*The United States Army Concept for Tactical Maneuver for the Future Modular Force: 2015-2024* 2006, 23).

While the first three aspects of this new concept apply quite precisely and, in some cases, more appropriately to an air-maneuver force, no mention was made in the document of the use of airpower as decisive maneuver. The Army's visionary move away from contact-centric warfare, however, is a giant step toward the capabilities an AMF brings to such a battle. An approach that builds situational awareness and maneuvers without closing for contact with the enemy in accordance with this new Army paradigm would be in perfect harmony with the BAO mission and BCA construct as well, which are both designed to optimize airpower in this precise scenario.

Similarly, the same manual made an encouraging reference to air maneuver without fully acknowledging its applicability to future brigade-oriented warfare:

> Informed by … the current focus on the BCT as the principal tactical fighting echelon, the primary means of maintaining continuous pressure in the future Modular Force will be division-directed commitment of maneuver BCTs in combined arms actions, while integrating near and deep fires with ground and air

> maneuver. Division commanders will sustain the continuity of their operations by cycling BCTs in and out of action and synchronizing logistical operations to that battle rhythm (*The United States Army Concept for Tactical Maneuver for the Future Modular Force: 2015-2024* 2006, 23).

By acknowledging airpower's capacity as a potential maneuver element, this document indicated the Army has moved forward in its concept development. The pamphlet also stated that "military history and recent operational history confirm the difficulty of maintaining high optempo in maneuver warfare when forces are organized with three or fewer maneuver units" (*The United States Army Concept for Tactical Maneuver for the Future Modular Force: 2015-2024* 2006, 22). The next step after recognizing air maneuver might be to accept that commanders could maintain or even increase this tempo by cycling an AMF interchangeably with Army BCTs. While requiring further development, such a transformational concept would be the acme of fully integrating interdependent air and ground forces in the Twenty-first century.

The Use of Ground Forces to Manipulate an Enemy

General Charles Krulak, former Commandant of the Marine Corps, remarked during his tenure that, "in the past, the goal was to use firepower to allow the infantry to close with and destroy the enemy with direct combat. Now we're thinking more of using maneuver not to close with an enemy … but to bring long-range fires on him. Maneuver will have more to do with staying away from the enemy than closing with him" (Blaker 1996, 21). During operations in which these long-range fires can be best delivered by airpower, ground forces will play a crucial role in bringing the enemy into the air-maneuver force's desired engagement zone. In order to target enemy ground forces most effectively, joint planners can employ land forces to manipulate and fix the enemy for precision air attack immediately prior to close ground combat. Because of surface maneuver's unique capability to influence an enemy, joint planners can use this strength to capitalize on the asymmetric advantage offered by BAO. According to *FM 3-0*, "maneuver creates and exposes enemy vulnerabilities to the massed effects of friendly combat power" (*FM 3-0* 2001, 6-13). Commanders who creatively employ ground forces to expose the enemy to air attack in this way will achieve a synergy that efficiently defeats a fielded force.

Experiences in Kosovo during the 1999 Operation Allied Force highlighted the importance of using a ground presence to keep the enemy reactive. Lambeth asserted that, "although Allied Force reconfirmed that friendly ground forces need no longer be inexorably committed to combat early, it also reconfirmed that airpower often cannot perform to its potential without a credible ground component in the campaign strategy" (Lambeth 2002, 17). Without the presence of a friendly ground force, enemy forces in Kosovo were able to disperse, conceal themselves, or shelter next to buildings, making air attack more difficult through such passive measures. If used by the JFC, a threatening or enticing ground force presence may have provided the motivation for the Serb ground forces to mass sooner and more frequently, forming a lucrative target for an air-maneuver force. In Afghanistan, this lesson was used to great advantage. U.S. Army Colonel John Antal remarked accordingly that "we saw the power of a truly joint air-ground attack. A single form of attack … would not have been as effective" (Antal 2002, 14).

Doctrinally, shaping operations may be designed to deceive the enemy, fix forces, or to force them to commit, and these shaping operations involve the use of maneuver forces to achieve desired effects (*FM 3-0* 2001, 7-9). For example, in support of decisive operations, a commander could fix an enemy force with a shaping operation as other forces envelop it in an attack against a decisive point (*FM 3-0* 2001, 7-7). Specifically, Army *FM 3-0* stated that commanders may direct a frontal attack, which is one form of maneuver warfare, as a shaping operation and another form of maneuver as the decisive operation (*FM 3-0* 2001, 7-15). When the joint commander uses ground maneuver forces to shape, but chooses to envelop with an air-maneuver force, he may achieve remarkably decisive results from the air well before the ground force is required to engage an enemy in a close fight. When airpower is applied as a maneuver force, these Army doctrinal concepts enable a commander to boldly plan to achieve decisive objectives while potentially sustaining a substantially reduced number of casualties.

When the JFC intentionally applies operational ground maneuver to channel the enemy into an AMF desired engagement zone, he optimizes joint effects. To prepare for this type of joint operation, the JFACC, in coordination with the land component commander, first determines the location of the optimum desired engagement zone based on terrain, order of battle, and subsequent objectives. Then, the enemy's position relative to friendly units will determine whether the land component should *push* or *pull* enemy forces toward that zone.

The "Hammer and Anvil" Approach

To *push* the enemy toward the desired AMF engagement zone, the friendly ground force should be of sufficient strength to threaten the enemy, thus leaving the adversary no other viable options. Furthermore, the opponent should have a clear channel toward the desired engagement zone, allowing the enemy to mass and become even more vulnerable to airpower. During this kind of offensive push, ground forces would serve as the anvil, and the JFC would use the air-maneuver force as the hammer. This transformational approach would differ from operations conducted in the past because, from the beginning, friendly surface forces would intend to slow or shift their forward movement at some point near the desired engagement zone in order to unleash the air maneuver concept. With troops avoiding close contact as much as possible, the JFACC would then apply the BAO concept to attack the massed enemy, without the need for inappropriate and inefficient close coordination measures. If circumstances inhibit the success of BAO, then the supported role would shift quickly to the land component and air support would be provided in a traditional CAS role. It should be mentioned that successful application of this approach would generally include an integrated deception plan and robust information operations.

Operation Anaconda, a large-scale ground operation conducted in the mountains of Afghanistan during Enduring Freedom, involved the movement of roughly 1,600 U.S. ground forces into the Shah-i-Kot Valley to capture or kill al Qaeda and Taliban fighters seeking refuge there. This operation, which would ultimately require the highest level of commitment from the air component during the entire war in Afghanistan, was fully planned, briefed and approved through General Tommy Franks, Commander of U.S. Central Command (CENTCOM) without any involvement of air component planners. Combined Joint Task Force (CJTF) Mountain, the organization established to plan and execute this mission, had no Air Force representation whatsoever during this planning period. Rebecca Grant found "the emerging plan for Anaconda had all the earmarks of an operation planned almost exclusively within the Army Component and special forces ... the plan for Anaconda had not been fully coordinated with the joint air component" (Grant 2002, 63). In fact, the final planning was complete and the operations order had been published before the air component commander, General Michael Moseley, or his staff had any knowledge about the ensuing operation (Lambeth 2005, 168-173).

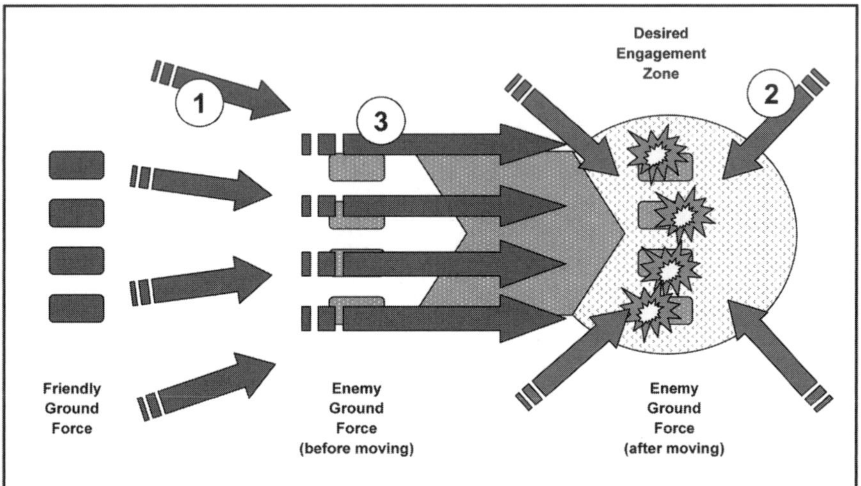

1. Friendly ground force serves as "anvil" that pushes enemy ground force into the desired engagement zone
2. Air maneuver force serves as "hammer" with ground force initially in support
3. BCT then attacks with airpower in support

Figure 4.1 Hammer and anvil strategy

This mission was a watershed event in the recent history of joint operations. Despite multiple examples of heroic execution on the part of the Soldiers and Airmen who participated in the operation, it represented a missed opportunity in joint operational planning and therefore incorporates valuable lessons. What Anaconda's planners overlooked was the chance to deliberately and intentionally plan to employ an air-maneuver force as the hammer. Although it was successful overall, an assessment of Anaconda's execution can provide insight into how much more effective this Coalition operation might have been had joint planners incorporated BAO in a pre-planned, integrated manner.

In February 2002, Coalition planners monitoring satellite and unmanned aerial vehicle sensors noted several hundred al Qaeda forces as they shifted from a retreating gaggle on the run into a concentrated mass of troops near Shah-i-Kot in the mountainous region of eastern Afghanistan (Grant 2002, 62). Taking stock of these fielded forces, CENTCOM initiated a plan to use Afghan forces in the north to push the enemy towards waiting U.S. forces that would encircle the enemy and block al Qaeda escape routes. This was a classic light infantry tactic, and from the outset, Operation Anaconda was predominantly an Army plan (Kaplan 2002, 23). In so doing, Army planners applied what might be viewed as a ground-only version "hammer and anvil" approach, with friendly Afghan troops as the anvil and U.S. Army ground forces serving as the hammer (Steele 2002, 19). Although Anaconda did not go as planned, the strategy was sound in that enemy forces reacted in a predictable way to ground-maneuver elements.

During the days preceding Anaconda, planners wisely chose to use ground maneuver forces to encircle and manipulate al Qaeda into positions vulnerable to the combined force but, regrettably, they failed to integrate the air component into the planning effort until very late in the process. Intending to achieve principal effects using troops on the ground, the joint planning team relegated the integration of air assets in Anaconda to a low priority, assuming that CAS aircraft would be available when needed, as a last resort. Army Major General Franklin Hagenback, who commanded Anaconda, explained that he chose to avoid the extensive use of airpower because "air campaigns are most effective against 'fixed' targets." The rugged terrain, caves, and fleeting forces were not compatible with this perception (Grant 2002, 68).

The operational concept for Anaconda, and the reason for its name, was to use American and Coalition Soldiers to squeeze the enemy from all sides of a 60 square-mile area as Soldiers located, fixed, and destroyed the enemy in rugged, mountainous terrain. Anaconda was hailed as the largest ground operation in Enduring Freedom, and the largest combat air assault in eleven years. In the middle of the AO was Objective Remington, a central zone towards which all planned assaults would drive (Steele 2002, 19). Army planners intended for several different ground units, including Afghan Coalition fighters, to either herd the al Qaeda forces towards vulnerable zones of engagement, or block them from escaping. The plan may have been sound but the performance of the Afghan troops upon which it hinged was less than optimum during execution. Nevertheless, as

General Franks reported, Afghan forces were able to move to contact with al Qaeda and, in some cases, actually served as blocking forces instead of driving the enemy forward. Eventually, the evading al Qaeda found their way into familiar prepared positions, which were pre-stocked with munitions dispersed within a series of cave systems. Although Coalition ground forces had effectively manipulated the enemy, they found themselves up against a formidable array of defensive fighting positions that made force-on-force combat a deadly option. Faced with a tough enemy, one U.S. detachment that had expected support from Afghan troops during a firefight instead called upon airpower to destroy a platoon-sized element (Haven 2002, 1).

Responding to the developing fight, the Coalition air operations center (CAOC) quickly directed a surge of airpower to the area, stacking assets in the airspace over the relatively small surface area in a queue to provide airpower to an engagement zone the size of the District of Columbia (Grant 2002, 67). As the land component pinned down enemy fielded forces, attack helicopters and air component assets delivered precise air strikes, inflicting heavy damage on al Qaeda. As General Hagenbeck reported, "we caught several hundred of them heading with RPGs [rocket-propelled grenades] and mortars toward the fight ... we body slammed them" (Grant 2002, 68). The battle subsequently enticed even more enemy forces towards the fight, with devastating consequences: Just two A-10 pilots were credited with more than 200 dead al Qaeda and Taliban (Grant 2002, 68). U.S. Air Force bombers, fighters, and gunships all contributed to ad hoc success during execution despite the absence of air component involvement during planning.

Although the outcome was acceptable, the preparation effort leading into the joint operation provided all Services useful lessons in joint campaign planning. During the days preceding Anaconda, planners' reliance on "last-minute" CAS, rather than detailed prior planning, ultimately forced Airmen to react to enemy movements while friendly units were in contact, rather than proactively engaging enemy forces on more favorable terms. Ground planners' fundamental misperceptions of air capabilities led them to limit application of airpower to fixed targets or CAS, missing the inherent flexibility offered by adequate numbers of pre-planned, laser-equipped strike aircraft on-call for precision BAO against inherently mobile enemy forces.

A plan developed in an integrated fashion by air and ground staff officers may have positioned friendly ground forces into the Shah-i-Kot area in a way that enticed a favorable response from the enemy, making them vulnerable to attack from the air. Rather than immediate close combat between opposing ground forces, exposed enemy combatants might have found themselves overmatched by Coalition airpower. When CJTF Mountain ground forces later closed for battle, they would likely have met with less resistance, fewer friendly casualties, and more definitive, rapid success. Rather than airpower rescuing friendly ground forces when they met with unanticipated resistance from the enemy, then, the air component might have fought alongside these ground forces from the outset in a fully integrated fashion.

Years earlier, during Desert Storm, Coalition planners inadvertently used the same hammer and anvil approach to create a successful outcome when U.S. Marines threatened to enter Kuwait City from the south, sending thousands of Iraqi troops in a massive retreat north towards the southeastern Iraqi city of Basra. As forces afloat used decoy maneuvers to deceive defenders along the coast, the 1st Marine Division battled Iraqi ground forces in southern Kuwait around the raging inferno of the Burqan oil fields. Despite a spirited Iraqi counterattack, the 1st Marine Division succeeded in overwhelming these enemy forces south of Kuwait City. This battle delayed a planned Marine advance into the city to face the occupying force. Nevertheless, the very presence of a threatening friendly ground force was sufficient to push the enemy north. In the evening immediately after the fighting at the Burqan oil fields, seismic sensors detected large-scale Iraqi troop movements, retreating out of Kuwait City towards Basra. What followed was one of the most destructive air attacks on a fielded force in history, as commanders dispatched U.S. Air Force aircrews to attack the moving column of vehicles in the dark, instructing them to stop the escaping convoy (Gordon and Trainor 1995, 370). The Airmen used infrared systems to target the lead vehicles in the convoy, halting the convoy in the desert at Mutlah Ridge. Air component aircraft continued the attack through the night, destroying 1,400 stolen vehicles and 28 Iraqi tanks (Gordon and Trainor 1995, 370). Coalition planners inadvertently used a formidable ground maneuver force to lead the enemy to respond in a way that made it vulnerable to attack from the air. Because of this, the Iraqi Army found itself on the horns of a dilemma: Stay in place and face an overwhelming ground force or move and risk debilitating air attack. They chose the latter, with predictably devastating consequences.

Enticement Strategy

Another method of enemy ground force manipulation occurs through enticement. This concept hinges on pulling rather than pushing the enemy. Using carefully coordinated maneuver, ground forces can bait an enemy, drawing it towards the desired engagement zone, while forcing it to coalesce and mass in a way most vulnerable to an air-maneuver force. When disbursed or entrenched units are threatened by Coalition surface maneuver, they must either remain in place, vulnerable to ground attack, or leave their shelters en masse to engage the attacking force and thus present themselves to air attack. With this knowledge, the JFC could direct operational surface maneuver in a way that forces adversary units to leave dispersed locations or protected underground facilities. Enemy forces may be drawn into the AMF desired engagement zone either because they perceive an advantage or see no viable alternative.

This approach has occurred on several occasions since 1990 with positive results and could very well be the most effective method of engaging troop concentrations hidden deep within underground facilities or other concealed positions. Although the ultimate outcomes were largely unintentional, surface units successfully enticed enemy formations into vulnerable air engagement zones in Desert Storm, Allied Force, and Iraqi Freedom. In all cases, surface maneuver drew the enemy into vulnerable positions where airpower, conducting what might now be called BAO, destroyed significant numbers of enemy forces.

During the first Gulf war, General Schwarzkopf's decision to utilize a Marine Expeditionary Force in a decoy role served to fix enemy fielded forces into positions vulnerable to attack along the coast. With the overall intention of weakening Iraqi ground forces prior to the allied ground assault, he directed Marines along the coast to practice amphibious operations to focus Iraqi attention towards the perceived threat from the Gulf. As General H. Norman Schwarzkopf reported during "The Mother of all Briefings", the allies then attacked Iraqi fortified positions from the air as they were held in place by surface forces (Gordon and Trainor 1995, 417). Although many defenders had already been routed by the time several Marine helicopter feints were executed, the Iraqis clearly anticipated an attack on the Kuwaiti coast and reacted accordingly (Gordon and Trainor 1995, 369). Marine ground maneuver divisions subsequently pinned them against the coast. The Marines afloat never engaged these forces, but manipulated them through decoy maneuver, fixing them in place. Although in this case, airpower wasn't used against these forces to the same decisive extent it was in other segments of Desert Storm, this successful decoy operation highlights the ability of a perceived ground maneuver force to generate a desired response from enemy fielded forces.

The Battle of Khafji, at the outset of Desert Storm hostilities, demonstrated in an irrefutable fashion, the unique ability of centrally controlled airpower to flexibly target enemy fielded forces as they are enticed into combat by friendly ground units. On 29 January 1991, Saddam Hussein directed a ground assault on the Saudi coastal town of Al Khafji. His apparent goal was to engage a joint force deployed along the coast, presumably to force an American ground battle that would produce casualties numerous enough to affect the political climate in America (Lambeth 2000, 121). An additional intent was to enable Iraqi troops to capture Coalition personnel (Palmer et al. 1998, 7). When reinforcing armored columns began to coalesce in Kuwait to support the initial Khafji invaders, JSTARS airborne sensors detected the vehicles, and Coalition fighter aircraft were immediately diverted to attack mobile targets in these columns. Lieutenent General Charles Horner, the JFACC during Desert Storm, then reassigned more than 140 jets against the Iraqi 3rd Armored and 5th Mechanized divisions, decimating the moving ground forces with deadly precision (Lambeth 2000, 123). Aircraft destroyed more than 350 tanks, 147 armored personnel carriers, and 89 artillery pieces during the ensuing several hours. This was the Iraqi military's last offensive ground operation. With overwhelming force, airpower prevented Iraqi reserves from reaching Khafji, inflicting heavy damage and forcing a hasty retreat (Palmer et al. 1998, 26). In *The General's War: The Inside Story of the Conflict in the Gulf*, Michael R. Gordon and General Bernard E. Trainor discussed an Iraqi officer who reported that his unit experienced more destruction by U.S. airpower in fifteen minutes than eight years during Iraq's war with Iran (Gordon and Trainor 1995, 286). Although Coalition troops were not intentionally positioned to channel Iraqi movement towards them, their presence caused the enemy to mass and move in a fashion vulnerable to airpower. When the battle was over, the inadvertent synergy created between a tempting ground presence and deadly airpower had led to remarkable success.

In 1999 during Operation Allied Force, airpower demonstrated an unprecedented capability to singularly achieve tangible results, despite a number of political constraints imposed by the Clinton administration and North Atlantic Treaty Organization (NATO) leadership. One such constraint was the decision not to use ground forces in the operation. As General Richard Hawley, former commander of Air Combat Command noted, the absence of friendly ground forces effectively eliminated the joint effects created by the coordinated use of land and air forces in an operation: "When you don't have that synergy, things take longer and they're harder" (Graham 1999). In hindsight, many believe that a credible ground force would have forced the Serbs to uncover and mass in ways which would have made them easier to find and attack from the air (Lambeth 2002, 17).

Despite this limitation, one particular Allied Force engagement underscored the remarkable synergy that can occur when ground maneuver units are used to manipulate enemy fielded forces into vulnerable air engagement zones. On 26 May 1999, the Kosovo Liberation Army (KLA), a guerrilla paramilitary force operating in Kosovo which opposed the Serbs, launched an offensive called "Operation Arrow" involving 400 troops against the Serbian forces in an attempt to secure safe supply routes for their fighters. Outgunned, the KLA rebels quickly found themselves on the defensive and were pinned down by over 700 massed Serbian troops. During these troop movements, Coalition sensor operators detected these Serbian concentrations and CAOC personnel subsequently directed unmanned aerial vehicles and fighter aircraft to the area for reconnaissance and attacks on these units (Lambeth 2000, 189). Later, on 7 June, Serbian forces staged an attack near the Kosovo-Albanian border against these KLA troops. In so doing, they exposed between 800 and 1,200 personnel to airpower in the attempt to overwhelm the guerilla fighters. As the KLA presence enticed these forces out of hiding, the JFACC directed two B-52 and two B-1 bombers using 86 blast fragmentation and cluster bombs to attack the Serbian ground force. Remarkably, battlefield reports after the engagement indicated that the Serbian offensive had encountered disastrous results and over 50 percent casualties. On that day, the synergy created by just four aircraft coupled with an enticing ground force resulted in an overwhelming defeat of the enemy. It was the heaviest and most successful Coalition attack against fielded forces during the 78-day campaign (Lambeth 2000, 190).

This unique engagement encapsulated the essence of the BAO mission advocated by this study. The aircraft flying in this battle were conducting neither CAS nor interdiction but something in between the two current counterland missions. Air planners applied an air-maneuver force on the battlefield against a fielded force, and soundly defeated the Serb unit before it was close enough to friendly forces to require the extensive positive control measures inherent with CAS. And most importantly, they did so before the enemy force could employ its own deadly weapons against the land force. Without airpower, the KLA guerillas would have likely suffered a crushing defeat. Likewise, without the KLA's presence, NATO airpower would likely have been unable to accomplish the task of rendering a substantial opposition ground force combat ineffective. Together, they dealt a crushing blow with minimal friendly losses.

Four years later during OIF, a column of between 70 and 120 Iraqi armored vehicles, under the expected cover of a heavy sandstorm, moved out of the city of Basra on 26 March 2003. Drawn by the presence of Coalition ground forces, and with the intent of engaging British troops in surface warfare, this convoy of troops and tanks streamed south, only to find themselves suddenly targeted by a massive assault from Harrier and Tornado attack aircraft, which effectively destroyed them (Richburg and Glasser 2003). Unintentionally, Coalition ground forces enticed an Iraqi surface unit into a BAO engagement zone with devastating results. A fundamental argument in this study is that planners, anticipating this kind of enemy reaction, can deliberately position surface forces to perform this role, intentionally exposing the enemy to air attack in the process. At the appropriate time, friendly ground units launch an offensive to decimate the beleaguered enemy force.

A few days later, as Army and Marine Corps forces moved swiftly north towards Baghdad, the Iraqi regime redirected troops towards them in an attempt to engage Coalition surface forces before they could enter the city. As reported in the *Washington Post*, "columns of Iraq's elite Republican Guard divisions and paramilitary guerillas moved south from Baghdad … under the shroud of

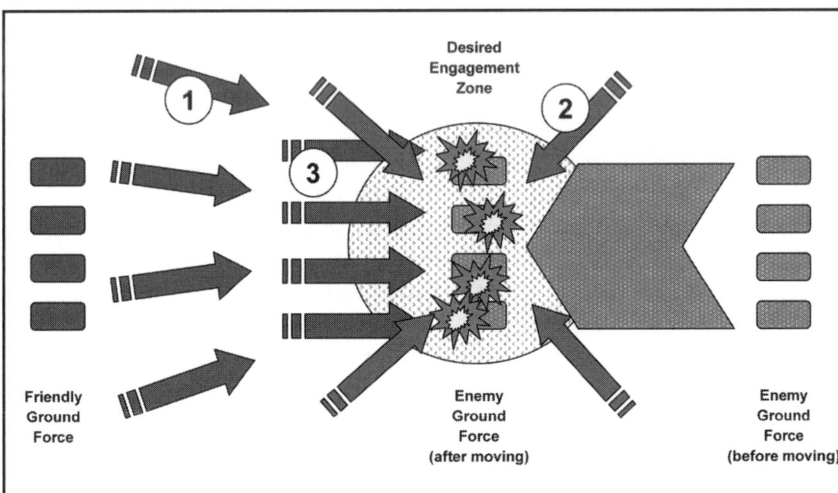

1. Friendly ground force serves as feint or fixing force that pulls enemy ground force into desired engagement zone
2. Air maneuver force attacks with ground force initially in support
3. BCT then attacks with airpower in support

Figure 4.2 Enticement strategy

swirling sandstorms in an apparent attempt to challenge [Coalition ground forces before they were] able to mass enough force to attack the capital …" (Baker and Chandrasekaran 2003, 1). Well prior to the arrival of friendly troops, Lieutenant General Michael Moseley, the JFACC during OIF, directed airpower against large concentrations of the Republican Guard arrayed in the region south of the city. "U.S. commanders responded by ordering intense air strikes, which they said wiped out much of one convoy of several hundred vehicles believed to be ferrying Soldiers from the Medina Division of the Republican Guard toward forward elements of the U.S. Army's 3rd Infantry Division encamped near Karbala, about 50 miles south of Baghdad" (Baker and Chandrasekaran 2003, 1). Responding to questions from reporters during a news conference after these operations, General Moseley remarked, "I find it interesting when folks say we're softening [the Republican Guard] up. We're not softening them up, we're *killing* them" (Scarborough 2003, 6). Echoing this, the *New York Times* explained, "the Republican Guard's resistance was so weak [because it had] been a bull's-eye for American airpower for days … [it had a] devastating effect on Iraq's forces in the field" (Gordon 2003). Some U.S. commanders, speaking with embedded reporters, expressed surprise at the lack of resistance that they saw as they rolled into Baghdad (Rumsfeld and Myers 2003). "People on TV kept asking, 'Where is the Republican Guard,'" said a fighter pilot who flew 19 combat missions in the war. "I can tell you where they are: They're blowed up. And we blew them up" (Trowbridge 2003, 8). In the final days of the Coalition drive to Baghdad, then, advancing troops enticed the movement of Iraqi forces into a zone vulnerable to overwhelming airpower, and the results were historic.

Feints and Fixing Forces

Because brigade combat teams are smaller and lighter than legacy armored divisions, future U.S. ground forces may find themselves facing numerically superior and possibly heavier enemy forces. The innovative application of airpower will be central to success in this new environment. Applying these past lessons in a more deliberate fashion, the joint force commander may choose to employ air as a maneuver force with the BCT initially serving as a feint or fixing force. In this role, the ground commander may position forces in a manner that is intentionally designed to force an enemy response favorable to friendly airpower. In response to the threat posed by the BCT, the enemy may establish hasty defensive positions. If it perceives an opportunity, the adversary's forces may attempt to attack. In so doing, these previously "dug in" forces would be vulnerable to attack from the air. In such a scenario, the BCT would not initially give battle, instead allowing airpower to "roll in" for the attack. Sun Tzu encouraged a similar approach a few thousand years ago: "All warfare is based on deception. Therefore, when capable, feign incapacity … Offer the enemy a bait to lure him; feign disorder and strike him … Pretend inferiority and encourage his arrogance" (Sun Tzu 1963, 66-67).

In accordance with his venerable teachings from millenniums past, the modern era's friendly ground force can lure the enemy as Sun Tzu directed and the air-maneuver force can strike him while he is unaware.

Additionally, this "bait and switch" approach will bring the traditional principle of mass into the Twenty-first century. Instead of *overwhelming* force, hallmark of the Twentieth century American way of war, the synergy achieved with air and ground forces operating in accordance with this transformational concept of air as a maneuver force will present the enemy with a vastly *overmatching* force. Once friendly surface forces assume the supported role and close for direct battle, the lethal BCT will crush the overmatched enemy. In a 2006 briefing presented at the Army War College, Air Force Colonel Roderick Zastrow compared the initial stage of this idea to pheasant hunting, where ground forces flush the enemy's fielded forces and airpower shoots them while they are on the fly (Zastrow 2006). Zielinski added clarity by further developing the same analogy:

> You could use air and ground forces in combination with air as the main effort in a situation where the enemy is dispersed or in caves. You send the ground force in to flush them out into the open. The friendly ground force acts like hunting dogs in a field full of quail. The dogs flush the quail into the air (out of their relative sanctuary). Airpower would shoot the quail as they attempt to flee or out-maneuver the ground forces (Zielinski 2006).

In a passage from its 2006 counterland document, Air Force doctrine acknowledged the utility of this approach to the AMF concept by offering two examples:

> The first involves airpower fixing enemy surface forces, thus allowing ground forces to engage. Airpower can hold enemy ground forces in place leaving friendly land forces free to maneuver. If the enemy counters surface maneuver with movement, losses from air attack (due to reduced concealment, greater detectability, and increased predictability) may become unacceptable. As a result, measures required to minimize losses from AI leave the enemy more susceptible to defeat by friendly surface forces. The second scheme involves surface forces fixing enemy forces, thus allowing airpower to engage the enemy. An actual or threatened surface advance can force an enemy to respond with counter maneuvers or resupply. By placing sustained pressure on the enemy, surface combat increases target acquisition by flushing the enemy from concealment thereby enabling airpower to destroy enemy forces at a faster rate than can be replaced. Close coordination among all components will help maximize enemy vulnerability to AI (*Counterland Operations* 2006, 29).

The document's reference to traditional AI terminology is notable. The concepts of BAO and BCA will greatly facilitate efforts to achieve this coordination in such a dynamic environment. Additionally, as discussed previously, these innovative approaches are arguably both required by interdependence and fully compatible with new ideas emerging in Army doctrine. As the Army looks to the future Objective Force, for example, it plans to use advanced reconnaissance to impact enemy fielded forces before initiating the close battle. Alan Vick and his co-authors explained this concept:

> Whereas, in the past, Army forces have usually sought contact with enemy forces, compelling them to reveal their positions and capabilities through combat, in the future, the Army hopes to attain a comprehensive picture of enemy forces *before* going into direct combat. Such a picture might allow the Army to defeat enemy forces with standoff and indirect-fire weapons rather than engage in more dangerous close-in fights (Vick et al. 2002, 10, emphasis in original).

Airpower has the inherent ability to develop such situational awareness from above and then attack these forces with a wide array of lethal weapons. Friendly forces positioned in ways quite similar to those envisioned in the Army Objective Force will likely elicit the response from the enemy's ground forces that will make them exceedingly vulnerable to air forces. These concepts, then, are compatible with and sometimes specifically mentioned in both Army and Air Force doctrine. Lacking is inter-Service integration and joint doctrine that fully acknowledges these promising ideas. Examples where the ideas were intentionally applied in actual combat operations are absent as well. The time is ripe for a detailed analysis of the AMF, BCA, and BAO concepts, synthesis of these ideas, and the application of them in the context of the Twenty-first century combat environment. Such is the purpose of the next chapter.

Chapter 5
Expanding *AirLandBattle21* Counterland Concepts

> However praiseworthy it may be to uphold tradition in the field of soldierly ethics, it is to be resisted in the field of military command. For today it is not only the business of commanders to think up new techniques which will destroy the value of the old; the potentialities of warfare are themselves being continually changed by technical advance. Thus the modern army commander … must be able to turn the whole structure of his thinking inside out.
>
> —Erwin Rommel

The counterland concepts advocated in this study are steeped in military theory. Sun Tzu's concept that balances the two forces of *cheng* and *ch'i* for maximum effect offers a remarkably pertinent example from the annals of martial history. In line with the popular Asian concept of *yin* and *yang*, Sun Tzu's *cheng* and *ch'i* lays a firm foundation for a practical illustration of how air maneuver theory may be applied in an actual battle.[1] By referring to a particular military force as *cheng*, Sun Tzu meant a normal, direct, regular, expected, or symmetric force. *Ch'i*, on the other hand, is translated as an extraordinary, indirect, irregular, unexpected, or asymmetric combat force. In *The Art of War*, Sun Tzu suggested that the sovereign should "generally, in battle, use the normal force to engage; use the extraordinary to win" (Sun Tzu 1963, 91). Samuel B. Griffith, in the introduction to his translation of Sun Tzu's text, offered that *cheng* and *ch'i* "are reciprocal; their effects are mutually reproductive. We may define the *cheng* element as fixing and the *ch'i* as flanking or encircling, or … as the force(s) of distraction and the force(s) of decision. Their blows are correlated" (Sun Tzu 1963, 42). Expounding on his translation, Griffith explained that "the normal (*cheng*) force fixes or distracts the enemy; the extraordinary (*ch'i*) forces act when and where their blows are not anticipated" (Sun Tzu 1963, 91).

This concept perfectly explains the potential relationship between a fixing ground force (*cheng*) and an air-maneuver force (*ch'i*). Parcel to Sun Tzu's concept

---

1 In 1979, a new convention for the phonetic spelling of Chinese characters was adopted. Because Samuel B. Griffin published his introduction to and translation of the *Art of War* in 1963, he used the traditional spelling of *ch'i* and *cheng*. The modern phonetic spelling of *cheng* is *zheng*. *Ch'i* is now *qi*. In order to avoid confusion and remain consistent with Griffith, who is cited repeatedly herein, the traditional spelling is used within this text.

of *cheng* and *ch'i* is the fact that the normal and extraordinary nature of these forces is not static but rather constantly flows and continuously shifts from one to the other. Like *yin* and *yang*, *cheng* becomes *ch'i* and back again in response to changing circumstances. According to Griffith, "the *cheng* and *ch'i* are compared to two interlocked rings: 'Who can tell where one begins and the other ends?' Their possible permutations are infinite; the *cheng* effort may be transformed into a *ch'i*, a *ch'i* into a *cheng*. Thus we may redefine a *ch'i* attack as one made where a decision is speedily attainable at least cost, in an area characterized by voids or fissures in the enemy's defences". Griffith further explained that "should the enemy perceive and respond to a *ch'i* maneuvre in such a manner as to neutralize it, the maneuvre would automatically become *cheng*" (Sun Tzu 1963, 42).

Again, this is precisely the essence of the relationship between air and ground forces in this application of *AirLandBattle21* counterland concepts. Employing air as a maneuver force places the enemy's fielded forces on the horns of a dilemma. If enemy forces react to ground action in a way that makes them vulnerable to airpower, airpower will attack the enemy in wholesale fashion. If instead the enemy chooses not to honor the threat posed by the friendly ground force (or takes subsequent action to shield itself from air attack that makes it vulnerable to ground forces), then airpower becomes the *cheng* force in support of ground *ch'i* maneuver. Sun Tzu observed later in his text that "he who knows the art of the direct and the indirect approach will be victorious. Such is the art of maneuvreing" (Sun Tzu 1963, 106). As in the times of Sun Tzu, the modern commander who recognizes the unique opportunities airpower brings as an indirect approach will arguably meet with the success Sun Tzu spoke of ages ago.

Like his Chinese predecessor, Mao Tse Tung similarly embraced a military strategy that dared "boldly to lure the enemy troop in deep so as to herd them together and annihilate them" (Mao Tse Tung 1975, 206). The concept of giving ground in order to make the enemy vulnerable to attack is well founded in military history. Mao, who had no airpower at his disposal, fully recognized its value: "We all know that when two boxers fight, the clever boxer usually gives a little ground at first, while the foolish one rushes in furiously and uses up all his resources at the very start, and in the end he is often beaten by the man who has given ground" (Mao Tse Tung 1975, 211). Accordingly, a friendly ground force applying this concept may willfully draw the enemy into an area most advantageous to the AMF poised for attack. Sun Tzu advocated exactly the same approach:

> Generally, he who occupies the field of battle first and awaits his enemy is at ease; he who comes later to the scene and rushes into the fight is weary. And therefore those skilled in war bring the enemy to the field of battle and are not brought there by him. One able to make the enemy come of his own accord does so by offering him some advantage. And one able to prevent him from coming does so by hurting him (Sun Tzu 1963, 96).

Bypassed Forces

During the major combat portion of OIF, many Iraqi Soldiers from bypassed, disintegrated or defeated units simply replaced their uniforms for civilian clothes and blended into the local population. Many of the same individuals later fought U.S. forces as terrorists, insurgents, and members of armed sectarian militias. As the Army develops plans to attack widely dispersed objectives with distributed, autonomous brigade-size units, bypassed forces will become an increasingly significant issue. The 2004 *National Military Strategy (NMS)* acknowledged the likelihood of this eventuality in Twenty-first century conflicts:

> The United States will conduct operations in widely diverse locations—from densely populated urban areas located in littoral regions to remote, inhospitable and austere locations. Military operations in this complex environment may be dramatically different than the high intensity combat missions for which U.S. forces routinely train. While U.S. Armed Forces' will continue to emphasize precision, speed, lethality and distributed operations, commanders must expect and plan for the possibility that their operations will produce unintended 2nd- and 3rd-order effects. For example, U.S. forces can precisely locate, track, and destroy discrete targets to reduce collateral damage and conclude operations as quickly as possible. Operations that rely on precision may result in large elements of an adversary's military remaining intact and segments of the population unaffected. Commanders must prepare to operate in regions where pockets of resistance remain and there exists the potential for continued combat operations amidst a large number of noncombatants (*NMS* 2004, 5).

This eventuality is precisely what the AMF concept is designed to address. Airpower is well suited to perform a variety of missions associated with such forces. For example, the air component could serve as the fixing force, allowing the BCT to flank an enemy unit by completely bypassing them as these highly mobile friendly ground forces directly engage and rapidly seize key enemy objectives. Any attempt from enemy ground forces to interfere with such BCT operations or logistical supply lines would be met with stiff resistance from the air. At some point, the air component may be assigned the mission of attacking, attriting, or neutralizing these bypassed forces. The BCT role during this portion of the operation may be to perform actions that elicit a favorable enemy response, making large numbers of bypassed forces vulnerable to attack from the air. As discussed, one of the many lessons from Operation Allied Force was that, in the absence of a credible friendly ground force, a thinking enemy will find multiple clever ways to conceal Soldiers and their equipment from air attack. Once forces from the KLA required the Serbs to respond to the threat of attack on the ground, the enemy became more vulnerable to airpower. In the same way, BCT actions intentionally designed to elicit such a response would place the enemy in quite a conundrum.

Dynamic Supported/Supporting Relationship

Central to the concept of an air-maneuver force is the supported/supporting relationship between air and ground forces. In operations involving the two, the prevailing assumption has traditionally placed the air component in a supporting role. The increased interdependence of the BCT has not only validated the importance of air in this supporting role but has arguably increased the criticality of the close air support that embodies it. Equally crucial, however, is the expanded airpower role that goes beyond traditional CAS responsibilities. The BAO mission and the associated AMF concept enable airpower, when appropriate, to serve in a supported role as the main effort. Accordingly, the 2003 *Air Force Basic Doctrine* document accurately argued that "air and space power is a maneuver element in its own right, coequal with land and maritime power; as such, it is no longer merely a supporting force to surface combat. As a maneuver element, it can be supported by surface forces in attaining its assigned objectives" (*Air Force Basic Doctrine* 2003, 16). More to the point, Air Force doctrine states unequivocally that, "in cases where air and space forces present the joint force commander (JFC) with the preponderance of counter-surface effects, it may be appropriate for the air and space component commander to be the supported commander for the destruction of enemy surface forces, with friendly surface force commanders acting in a supporting role" (*Operations and Organizations* 2006, 4).

The idea of the air component assuming the supported role for missions such as intelligençe, surveillance, and reconnaissance (ISR) and strategic attack is generally well accepted and captured in current doctrine. More novel and contentious is the concept that, when integrated air and ground forces attack the enemy's fielded forces, the friendly ground force would support the air-maneuver force. At first glance, the idea makes good sense. As Army Colonel De Oliveira conceded, "my opinion is that whatever unit can achieve the commander's intent should do so and any other units required should support that unit" (Oliveira 2006). Air Force doctrine arrived at a similar conclusion in 2006: "In some circumstances ground maneuver may support aerial maneuver by forcing the enemy into a position that is more vulnerable to air attack, enabling air and space power to deliver a vital blow" (*Counterland Operations* 2006, 10-11).

When theory meets practice, however, the concept gets a bit tricky. Ceding the supported role to an air element unable to hold ground might be difficult for some ground commanders. Air Force doctrine discussed a possible source of this disconnect:

> From a Soldier's perspective, shaping operations support the decisive operation by affecting enemy capabilities and forces, or by influencing enemy decisions. Shaping operations use all elements of combat power to neutralize or reduce enemy capabilities ... As a result, Soldiers may consider AI as shaping which solely supports their maneuver elements. From an Airman's perspective, AI may be conducted either in support of surface force objectives or in direct support

> of JFC objectives; in the latter case, the air component commander might be the supported commander. Because of these slightly differing views, there is a potential for friction between the air and land components regarding supporting/supported roles and responsibility for planning. These situations require careful and continuing dialogue between the competing senior commanders and their common superior commander (*Counterland Operations* 2006, 19).

Replete with reference to air maneuver, emerging Army doctrine still makes no reference to ground forces supporting a main effort from the air. The Air Force perspective, on the contrary, clearly carves out a role for airpower well beyond the traditional support function: "The Airman's perception of depth differs from that of the Soldier, in that air and space power can reach to any depth of the battlespace—from the close battle area back to and beyond the enemy's heartland. As an aerial maneuver force, counterland operations forces should not be considered as 'flying artillery'" (*Counterland Operations* 2006, 10). The pragmatic examples, newly defined missions, and practical operating concepts offered in this text will help move the U.S. military toward future application of the air maneuver concept during actual combat operations.

One impediment to further progress in this area, however, is confusion over traditional terms as they are being applied in new ways. A discussion about the terms *fires* and *maneuver* in the 2006 *Counterland Operations* publication highlights the point. The document began by quoting the joint definition of fires as "the effects of lethal or nonlethal weapons" and *joint fires* as "fires produced during the employment of forces from two or more components in coordinated action toward a common objective." In the past, both Army and joint doctrine often equated airpower to fires. This Air Force document, on the contrary, drew an important distinction between the Air Force counterland mission and fires:

> Counterland itself is not joint fires; rather, it represents a form of aerial maneuver, which delivers fires on various targets as required. Those counterland missions that are apportioned to support another component, such as CAS and some AI, can be defined as meeting the description of "two or more components in coordinated action." Therefore, the application of these missions can be called joint fires. Those missions that operate in direct support of theater strategy, such as theater-wide AI, are not operating in "coordinated action" with another component, rather they are conducting missions with organic forces in support of a scheme of maneuver. Therefore, the fires produced by those missions are not joint fires (*Counterland Operations* 2006, 11).

Moving forward toward greater air and ground force integration requires mutual understanding of this terminology. The term *maneuver* has traditionally implied the main effort and the supported force. *Fires*, on the contrary, have served as the supporting effort and, more specifically, in support of maneuver. Similarly, feints or fixing forces, while making important contributions to the ground scheme of maneuver, were also viewed as supporting the main effort. While the term *fire support* was commonplace in the past, for example, a term such as *maneuver support* would have been counterintuitive and consequently does not exist.

The worthwhile effort to adopt these terms into Air Force lexicon may have introduced some divergence in their meaning. The following passage from *Air Force Basic Doctrine* illuminates the point: "Counterland provides the JFC two discrete air operations for engaging enemy land forces: air interdiction (AI), in which air maneuver indirectly supports land maneuver or directly supports an air scheme of maneuver, and close air support (CAS), in which air maneuver directly supports land maneuver" (*Air Force Basic Doctrine* 2003, 44). As Air Force doctrine continues to change and begins to fully embrace Army concepts in the Twenty-first century, two aspects of this viewpoint are ripe for adjustment. First is the counterintuitive suggestion that one maneuver force can support another. As the names suggests, CAS supports land maneuver. Consequently, airpower does not serve in a traditional maneuver role while performing the CAS mission. Similarly, while serving in the AI role, airpower can either support or serve as the maneuver force, but cannot do both at the same time. The other area inviting revision in current Air Force doctrine is the assignment of only two options for the counterland mission. Airpower is clearly supporting in the CAS role and doctrinally supported when conducting strategic attack. Typical of the overly broad AI mission, however, the supported/supporting relationship is equally vague when applied to air interdiction. Consequently, a current source of confusion lies in the absence of a clearly defined counterland mission that places an air-maneuver force in a supported role. The addition of BAO to the counterland suite would add clarity and greatly improve mutual understanding between the Services in this area. One method to further enhance inter-Service planning and execution of joint operational concepts may be to emphasize integration, rather than focus purely on the supported/supporting relationship.

An example of integration, from the Airman's perspective, illustrates flexible roles within a framework of dynamic options, selected by an airborne on-scene commander, the two-ship flight lead. Air Force fighter pilots seldom fight as a single aircraft. Instead, they are taught to fly and fight in a tactical combat formation. The building block of any strike package, regardless of size, is typically the two-ship formation. At the core of this tactical formation is the concept of mutual support, where each aircraft is dependent on the other for survival. The traditional employment position features the two aircraft line abreast so that each can "check six" for the other. Each wingman ensures that no enemy aircraft can move unobserved into an attack position behind the other. In this line-abreast formation, both wingmen are also in an ideal position to attack an enemy fighter when the opportunity presents itself. When one wingman identifies an enemy aircraft, he becomes "engaged" in the attack and, with a single radio call, the other shifts to a supporting role. This supported/supporting construct is remarkably dynamic and, in the event that the supporting wingman later becomes engaged, a simple radio call can change relationship once again. The two-ship formation provides a useful analogy for the interdependence of air and ground forces employed in accordance with the concept of air as a maneuver force. Each can serve interchangeably in supported or supporting roles and, in so doing, each increases the effectiveness of the other's attack. Furthermore, either can attack while the other serves in a supporting role. If employed in accordance with the concepts furthered in this study, the supported/supporting relationship can be quite dynamic, allowing air and ground forces to shift roles when perceived opportunities allow or threats require.

As is often the case, however, the devil is in the details. Critics of the concept might find shifting the supported/supporting relationship between air and ground forces far too challenging in a dynamic combat environment. Interestingly, while not prevalent in joint or any other Service documents, Marine Corps doctrine fully accounts for the feasibility of air serving in a supported role and recognizes the importance of shifting the supported-supporting relationship in an expeditious manner. The 2001 Marine Corps Doctrine Publication (MCDP) 1-0, *Marine Corps Operations*, discussed the potentially dynamic relationship between the main and supporting efforts: "While a commander always designates a main effort, it may shift—either planned or unplanned—during the course of battle as events unfold. Because events and the enemy are unpredictable, few battles flow exactly as the commander has planned." As such, the Marine Corps document found that "a supporting effort may achieve unexpected success during execution. As a result, the commander must make adjustments. After assessing the changing situation, he may designate the supporting effort as the main effort" (*Marine Corps Operations* 2001, 6-26). More to the point, though, this relationship does not relate only to Marine ground units but may be applicable to the aviation combat element (ACE) of the MAGTF as well: "The flexibility inherent in Marine aviation allows the commander to shift the main effort to the ACE rapidly and usually with significantly less repositioning of resources than other forces of the MAGTF" (*Marine Corps Operations* 2001, 6-27). It is precisely this flexibility and relative ease of shifting the supported/supporting relationship between air and ground forces that makes the concept of air as a maneuver force such a viable option in Twenty-first century U.S. warfare.

Designating air as the supported component during portions of major combat operations will emphasize coordination of joint three-dimensional maneuver warfare because it will, by default, force thorough, well-planned integration of both land and air capabilities into the operational planning effort. Operation Anaconda demonstrated the unfortunate outcome of using insufficiently planned integration of air and land component capabilities. If Anaconda indicates typical planning constructs, a planning staff with predominantly ground-oriented planners may not be fully equipped to develop an AMF strategy featuring ground units in a supporting role. Similarly, a staff of air planners may not possess an adequately broad and thorough understanding of ground operations. Therefore, because operations based on this construct will likely involve substantial ground and air operations, planners in both uniforms must thoroughly collaborate to identify desired engagement zones, direct the manipulation of enemy forces, and to pre-coordinate for a timely exchange of the supported/supporting roles. In the future, operations requiring Anaconda's hammer-and-anvil style of warfare could be fought more efficiently using this construct. Such operations could be planned and executed more effectively as well by a truly joint team of air and ground strategists performing their duties while collocated in a joint operations center or a joint warfighting headquarters.

At a given point in a particular operation, friendly ground forces will shift from a manipulation mission to actively engage enemy forces, demanding a transition of the supported role to the land component. The timing of this shift will depend on factors such as the coercive ability of the ground maneuver force to compel the enemy towards the air component's desired engagement zone. Under ideal circumstances, if the ground force successfully entices a favorable enemy response, the air-maneuver force will likely reduce the enemy force strength significantly, allowing ground forces to achieve JFC objectives with reduced casualties. In the event that the air component is unable to conduct meaningful BAO against fielded forces, friendly ground forces would assume the supported role and airpower would revert to more traditional roles.

Because of the dynamic nature of this proposed style of operation, the speed, simplicity, and clarity of the role reversal between components must be a top planning priority. Communication will be crucial, and commanders at all levels of organization must understand the plan in order to achieve the desired results. As James Winnefeld observed, "the joint commander's objective should be to minimize the complexity while maximizing the product of the joint force" (Winnefeld and Johnson 1993, 164).

The Vertical Flank

The Army's 2006 *Operational Maneuver* pamphlet presented an intriguing concept of "vertical maneuver" that "changes the geometry of the battlespace" and "enables distributed operations within a non-linear battlespace." This transformational, three-dimensional view of the battlespace offers a new paradigm for Twenty-first century warfare. Unfortunately, as currently written, the concept seemed to exclusively apply to the employment of Army ground forces with no specified tie to airpower beyond the airlift required to enable the approach. To be fair, the document, which did make reference to "simultaneous engagement of air-ground maneuver elements", is unquestionably a step in the right direction (*The United States Army's Operating Concept for Operational Maneuver: 2015-2024* 2006, 24-25). This citation, however, would generally be understood as referring to indigenous Army aviation assets such as attack helicopters. David Fastabend expanded on the emerging Army concept:

One of our more ambitious expectations for future operational art is intra-theater operational maneuver, that is, the ability to vault over or around enemy dispositions rather than through them and doing so with mounted, rather than dismounted, forces. Historically, with a few exceptions (Inchon), the best examples of operational maneuver first involved major tactical breakthroughs that permitted and led to exploitation and operational envelopments. If we can achieve the routine application of three-dimensional operational maneuver, we will routinely alter the geometry of the battle space with concomitant impacts on operational art (Fastabend 2005, 162).

Similarly, Fastabend discussed three-dimensional maneuver without reference to combat airpower.

Traditionally, Soldiers have viewed the battlefield as two dimensional and accordingly acknowledged the existence of flanks on each side of an advancing Army. The advent of airpower, however, gave birth to an additional flank from above (see Givens 2002). Under the title "Air and Space Maneuver Warfare", 2006 Air Force doctrine makes reference to the concept of a vertical flank:

Air and space power's vertical position within the theater of operations confers upon it a distinct advantage. Traditionally, the physical structure of ground maneuver forces has consisted of fronts, flanks, and rears. Although these terms have no application to air and space power, it is sometimes useful to make an analogy in surface terms in order to convey air and space power's contribution. In such terms, airpower adds a flank in the third dimension that makes the vertical battle as important as the horizontal battle. Using the metaphor of surface flanks, the airspace above the theater is also maneuver area but in three dimensions, and can be taken and exploited to either destroy or dislodge an enemy or achieve a relative advantage. Thus, as with surface flanks, a commander should seek to gain a position of advantage by turning an enemy's vertical flank, and should no sooner expose one's own vertical flank (*Operations and Organizations* 2006, 3).

The inherent vulnerability for an armed force unprotected in the vertical can, by the same token, be a tremendous opportunity for a visionary opponent who chooses to seize it. As the Army transforms into a highly mobile force, it has begun to fully grasp the great utility of the third dimension. Operating concepts are now literally elevating ground forces in this new direction. The next step in this journey toward full transformation is to recognize the inherent and natural capacity for an air-maneuver force to turn the enemy's vertical flank. In the same manner, graduate-level integration may explore ways to simultaneously employ air and ground forces in this domain. One example might feature ground forces employing vertical maneuver to fix the enemy as airpower attacks from above in ways similar to the approaches discussed in the previous chapter.

Airpower in Traditional Ground Roles

The increased interdependence of air and ground forces will encourage future planners to explore opportunities for airpower to perform missions traditionally performed by ground forces. Non-contiguous BCT employment methods, as an example, will make it increasingly necessary for airpower to look beyond the vertical flank and protect a brigade's flanks in a traditional, horizontal sense. Because BCTs will operate at times fully separated from other ground forces, air forces will have to provide 360-degree protection, guarding in a sense a "flank" that fully encircles the BCT. In Army Major Jason Mead's view, the non-contiguous nature of the Twenty-first century battlespace places a high priority on airpower's capacity to protect supply lines and critical lines of communication (LOCs) between the areas of operation (AOs) of autonomous, geographically separated BCTs. He explained the importance of airpower providing "more coverage of ground LOCs for logistical units caught in the 'no man's land' between BCT AOs, where division and corps have no assets to support." Mead went on to explain that, in OIF, "as a company commander for a company that conducted logistics patrols, we were heavily reliant on air assets … as well as fixed-wing aircraft as the ultimate 'QRF' [quick reaction force] whenever we got into troubles." He concluded by submitting that "integration of air power as the combat power and logistic convoys as the eyes on the ground have an excellent chance of success, if they can be coordinated" (Mead 2006). This concept is as applicable to today's fight in Iraq as it is to tomorrow's wars elsewhere. The bypassed forces, so likely present in the future BCT non-linear fight, may challenge supply lines in a conventional way as they have unconventionally in Iraq. Airpower is ideally suited to address this predictable issue. The JFC, for example, can use mission-type orders to direct the air component to isolate a particular region from the fight, neutralize enemy forces in this region, and prevent these bypassed forces from affecting the supply lines supporting geographically separated BCTs.

Fully integrating air and ground forces as advocated in this study opens the door to literally limitless opportunities for the use of airpower in roles traditionally held by ground forces. Curry, for instance, envisioned "the Air Force responsibility of monitoring all traffic running between major population centers using you guys [Air Force assets] in a traditional cavalry mission" (Curry 2006). Similarly, Johnson posited that "the BCT will not likely be as mobile as desired." Because of this projected limitation, he identified another potential airpower mission compatible with the precepts of *AirLandBattle21*: "A need may emerge for a force to be able to protect a position from something like a 'Second Echelon' attack, a force whose strength is beyond the capability of an already engaged BCT" (Johnson 2006). This reference to the basic thrust of Soviet-era AirLand Battle keenly demonstrates the increased dependence of Twenty-first century ground forces on air assets and validates the need for concepts embodied in *AirLandBattle21*.

Helicopters in Iraq

Some may associate the AMF concept and related ideas for employing airpower in nontraditional roles to the use of Army helicopters in a similar fashion. Hymes, for instance, argued that, "in the Army, an Aviation Brigade has always been able to operate as a Main Effort and other efforts operate as supporting efforts. I really don't see a change to this concept" (Hymes 2006). Likewise, Curry acknowledged the feasibility of fighter and attack aircraft serving as what he referred to as "high-speed cavalry." He submitted that "Army aviation has been fighting this for fifteen years. We've gone back and forth on Army aviation as a maneuver element. In some ways, we've gained a lot of traction on that but then we proceeded to oversell our capabilities and that's the problem." The challenges encountered by V Corps' 11th Attack Helicopter Regiment (AHR) during OIF may provide an illuminating example of Curry's point. In fact, Johnson explained that, "as things begin to emerge from the shadows of ideas and experiments, I expect we will see that the inability of the BCT to become nearly as mobile as hoped will militate rather strongly FOR air as a maneuver force. Those opposed will cite the example of the 11th Aviation Brigade's abortive attempt to act as a maneuver/strike force during OIF" (Johnson 2006). The ensuing analysis of failures during this operation will argue persuasively that the AMF concept is sound but that the asset selected may not have been a good fit for the mission.

On the night of 23 March 2003, 31 Army Apaches attacked the Iraqi Medina Division positioned in the vicinity of the Karbala Gap south of Baghdad. Murray and Scales explained that Lieutenant General William Scott Wallace, the commander of V Corps, and Major General Buford "Buff" Blount, 3rd Infantry Division (ID) commander, "recognized the looming battle that was coming with the Medina Division and other Republican Guard formations. Both believed that Iraqi defenders needed to be attacked from the air before the 3rd Infantry Division reached Baghdad's defenses" (Murray and Scales 2003, 105). In fact, Wallace submitted that "the Medina is the key to this whole thing … our focus was always on the Medina Division, for twelve months" (Atkinson 2004, 67). In the absence of a clearly articulated Twenty-first century operating concept that fit this scenario, they resurrected familiar concepts from the past: "In effect they were drawing on the army's Cold War concepts of Deep Battle and AirLand Battle … The deep attack plan called for the air force to wear the Medina down with precision weaponry" (Atkinson 2004, 67). This scenario, then, seemed a perfect fit for an *AirLandBattle21*-type approach using fixed-wing aviation assets from the air component. Wallace and Blount, however, "worried that this division would be the most difficult to find and destroy with high-performance aircraft. Intelligence indicated that the Medina had skillfully dispersed its fighting elements within the suburbs and villages lying west of Baghdad." As such, the ground commanders passed on the opportunity to offer the mission to the air component, concluding that "the only way to neutralize the Medina would be to attack it with low flying helicopters. Wallace gave the mission to the 11th Attack Helicopter Regiment

(AHR), which was directly assigned to V Corps" (Murray and Scales 2003, 105). In light of the ground commanders' concerns about fixed-wing capacity to target concealed ground forces, positioning a 3rd ID BCT in a particularly provocative manner may have forced dispersed Iraqi units to react in ways that made them vulnerable to attack from the air component. This option, however, was not considered.

Expectations were high. "When we attack the 14th Brigade of the Medina Division", said Colonel Thomas J. Schoenbeck, 101st Airborne Division (Air Assault) Chief of Staff, "we expect to see a hundred burning armored vehicles within half an hour" (Atkinson 2004, 124). Wallace explained, "I want to destroy the Medina and let the Hammurabi and other Republican Guard divisions think about that. We're going to take our time, we're going to kill every one of them, and we're not going to move forward until we do" (Atkinson 2004, 124). Seemingly ignorant of these lofty expectations, though, Iraqi observers employed rudimentary methods, such as cycling the city's electrical power, and used cell phones to alert fellow defenders as the helicopter formations approached the target area. One Apache pilot reported the results: "Large, *very* large-caliber triple-A started coming at us." In fact, the slow, low-flying aircraft were greeted by various weapons effects which included not just anti-aircraft artillery but small arms fire from rifles and machine guns as well as rocket-propelled grenades. By another account, "the helicopters flew into a wall of lead" (Atkinson 2004, 149-150). According to one commander, "there were eight or ten heavy streams of fire coming out of the city ... we never really had contact with enemy tanks or artillery" (Atkinson 2004, 150). In fact, Atkinson reported that "not many pilots from the 11th got close enough to engage the Medina; one squadron aborted before ever reaching enemy entrenchments. Few Hellfire missiles were fired" (Atkinson 2004, 152). Regrettably, despite the results Army planners anticipated, 11th AHR Apaches damaged a total of only six armored vehicles and three trucks. To make matters worse, one Apache crashed on takeoff, another was shot down during the attack, and a pilot in a third helicopter was shot through the neck (Atkinson 2004, 136). Twenty-eight of the 29 helicopters that returned "sustained some damage" with an average of fifteen to twenty bullet holes in each aircraft. Battle damage required the replacement of six engines, 16 main rotor blades, five drive shafts, and six tail blades. The regiment would not be fully mission capable for a month following the attack (Atkinson 2004, 153).

During Operation Allied Force in 1999, the debacle with the Army Apaches of Task Force Hawk drove the Army Chief of Staff, General Shinseki, to begin Army transformation. Four years later, with transformation well on its way, the absence of transformational concepts for the use of the same Army assets led to similarly discouraging results. The bravery and tenacity of the Apache pilots was not in question. What may have invited scrutiny, though, was the decision to assign helicopters to a mission arguably more appropriate for fast-moving, high-flying, and more survivable fixed-wing combat aircraft. The Army aviation regiment's troubles were not an indictment of the air-maneuver force concept.

Instead, they constitute a case where the Apache, while exceedingly capable when appropriately tasked, was simply not the right platform for aerial attacks in a relatively well-defended suburban area. Selecting more survivable aviation assets readily available in the air component may very well have yielded a different and more favorable result. Fully recognizing the Twenty-first century capabilities of fixed-wing airpower and applying transformational ideas for its application to the ground battle may drastically alter the outcome of future operations.

In the aftermath of the 11th AHR mission, the 101st Airborne Division aviation brigade conducted a similar attack five days later. According to Murray and Scales, Lieutenant General Wallace, V Corps commander in charge of the operation, "knew that a ground force could not maintain the speed of its advance without some means to protect its rapidly lengthening supply line" (Murray and Scales 2003, 98). Once again, such a scenario constituted the ideal opportunity for employing an AMF in place of a BCT to enable continued high operational tempo. Wallace, however, initially picked the 101st Airborne Division for the task. While the aircraft sustained less damage, RAND Researchers John Gordon, David Johnson, and Peter Wilson reported that the strike exacted little damage on Iraqi forces and explained that, due to the vulnerability and consequent ineffectiveness of their attack helicopters, the Army "attempted no more deep attacks against regular Iraqi Army formations" (Gordon et al. 2007, 68). As the AMF concept and other precepts of *AirLandBattle21* become more widely known, understood, and accepted, future ground commanders may more naturally opt for fixed-wing air maneuver when it constitutes the best force for the job.

In preparation for this second large-scale helicopter strike, the division conducted what it reported as the longest air assault in U.S. Army history (Atkinson 2004, 135-136). This 400 kilometer trek from Udairi Airfield in Kuwait was conducted for the purpose of establishing forward area refueling points (FARPs) for ingressing attack helicopters (Atkinson 2004, 135-136). In the words of General David H. Petraeus, then commander of the 101st Airborne Division, "the purpose of this division, initially, is to position the Apaches to get into the fight … Apaches are to be the star of the show" (Atkinson 2004, 66). This approach was reminiscent of the Allied "island-hopping" campaign during the Second World War that deliberately seized islands in an incremental fashion in order to establish airfields closer to Japan. Interestingly, then, Army leaders displayed a willingness to conduct distributed ground operations in support of an aviation mission. While certainly a step in the right direction, it may have once again been more appropriate to conduct such a mission in support of an air-maneuver force consisting of more survivable fixed-wing aviation resident in the air component. Army commanders and their planners may not have fully acknowledged the tremendous contributions fixed-wing air component assets had already made to the fight in Iraq as well as the critical impact such aircraft could have continued to make had the JFC assigned this particular mission to the JFACC. Of note, the longer ranges and greater loiter times of fixed-wing combat aircraft paired with their ability to air refuel at high altitudes, well removed from the immediate battle, negate the need for ground-based FARPs. Rather than conducting risky missions to seize terrain for this purpose, then, future BCTs could instead support an AMF in the many far more transformational ways discussed in this chapter.

Beyond Independent and Supporting Airpower

In the past century, airpower was traditionally viewed in two distinct ways. One perspective commonly espoused by advocates envisioned airpower as operating separate from land forces. The essence of this approach was the belief that attacking strategic targets could win the war independent of ground forces. Because they saw this capability as making the U.S. Air Force unique, those sharing this approach would offer strategic attack as the Service's central and defining mission. The other prevailing mindset envisioned airpower predominantly in the role of support to ground forces. This view, held by many in the U.S. Army, most often pointed to "boots on the ground" as the only dependable way to win wars, relegating airpower to its traditional supporting role.

In the Twentieth century, then, the Air Force often advocated what might be called "air as an *independent* force." A basic premise of this study is the recommendation that, in the Twenty-first century, the Service moves beyond this past approach and instead adopts the new paradigm of air as a maneuver force. Additionally, the current transformation mandates an innovative view of air and ground forces as *interdependent*—vice independent—and the recognition that air forces can and, in some cases, should be supported by ground forces instead of the other way around. Fully integrating formerly independent or deconflicted air and land forces in this way, while allowing for a dynamic supported-supporting relationship as they shift roles during future battles, opens the door to a variety of necessary options not commonly explored.

An illuminating passage from the 2003 *Air Force Basic Doctrine* document made reference to the use of airpower in both support and independent roles:

> Although historically associated with support to friendly surface forces, counterland operations may encompass the identical missions, either without the presence of friendly surface forces or with only small numbers of surface forces providing target cueing. This independent or direct attack of adversary surface operations by air and space forces is the key to success when seizing the initiative during early phases of a conflict (*Air Force Basic Doctrine* 2003, 43-44).

Interestingly, though, the only two options this doctrinal explanation of counterland offers were air operations either directly in support of or fully independent of friendly ground forces. Not mentioned in this bipolar view, however, was the transformational approach of employing the same air and ground forces *interdependently* with each playing supporting roles interchangeably. The traditional Army approach, then, viewed airpower in support of ground forces. The Air Force historically explored roles such as strategic attack where airpower operated independent of ground forces. *AirLandBattle21* is novel in that it goes beyond these independent roles, recognizing the quintessential interdependence between Twenty-first century air and ground forces, and expands the limited view of air as support, allowing for a dynamic supported/supporting relationship between these forces.

Interdependent Air and Ground Maneuver

Warfare in recent years has demonstrated how land and air components can be used synergistically to produce dramatic effects on the battlefield. While ground maneuver can influence an enemy, either to push him away under threat of defeat, or to entice him to fight, hoping for victory, this important capability becomes most effective when combined with an air-maneuver force. What has begun to emerge is the potential ability for a joint commander to *intentionally* manipulate an enemy ground force into a zone vulnerable to massed, concentrated airpower. In the application of force against an Army, the Air Force's newest doctrinal construct, BAO, offers the JFC a methodology that ushers in a new era for joint operational planning. The mutually supporting concepts of BAO, AMF and BCAs capitalize on lessons learned during operations in Kuwait, Kosovo, Afghanistan, and Iraq, where enemy forces reacted to a hostile ground presence, only to be subsequently hammered by airpower. These operations have demonstrated, more by consequence than design, that using BAO in a campaign against fielded forces, the JFC may best achieve his operational objectives by designating the air component the main effort, supported by surface maneuver to drive or entice an enemy into a vulnerable air attack zone, then shift the supported role to the land component as troops move to close contact. Incorporating these concepts into Air Force and joint doctrine will enable Twenty-first century planners to incorporate the remarkable lessons of recent warfare into future operations in a more deliberate fashion.[2]

Critics may argue that the concept of air as a maneuver force has been around for quite some time and is incorporated into both Air Force and Army doctrine. While not yet captured in joint doctrine, it is true that much has been written on the concept. There have been some articles advocating the complementary ideas of BAO and the BCA construct as well. Applying concepts, however, can be far more difficult then developing them. The real challenge with a concept as transformational as the air-maneuver force is moving from theory to practice and truly implementing it in the battlespace in a deliberate manner. The ideas presented in this study aim to enable the widespread use of the AMF in the Twenty-first century fight. In so doing, perhaps the joint team of the future will "bring home the gold" as the U.S. Olympic basketball program hopes to do repeatedly throughout the Twenty-first century.

---

2 See, Jinnette, J. (2003), *Employing an Air Maneuver Force: Battlefield Air Operations with Surface Maneuver in a Joint Campaign.*

Chapter 6
Counterair Concepts

> Victory smiles upon those who anticipate the changes in the character of war, not upon those who wait to adapt themselves after the changes occur.
>
> —Guilio Douhet

The idea of air superiority is inextricably linked to American airpower, growing up with U.S. aviation as it was becoming a viable military weapon. As American airpower became more dominant later in the Twentieth century, the concept of air supremacy became vogue. Today, the largely unchallenged preeminence of American airpower, as represented by the F-22 Raptor, has led to the advent of air dominance. In a large U.S. Air Force with plentiful fighter wings and a plethora of fighter aircraft, unrestricted control in the air may be possible. Current fiscal realities and resource limitations, however, bring the concepts of air supremacy and air dominance into sharper focus. Budgetary constraints, paired with the need to recapitalize an aging fleet, have left the Air Force little choice other than to reduce the number of aircraft and personnel. Current estimates indicate the Air Force will lose as many as 350 fighter aircraft in the near future. In their place will be roughly half the number of F-22s. Over seventeen years of sustained air operations in Southwest Asia has stretched expeditionary elements of the Air Force. Currently, the Service is fully engaged in Afghanistan, Iraq, and the wider war on terror. The Twenty-first century strategic landscape features numerous additional potential threats to U.S. national security across the spectrum of conflict. Meanwhile, the Army is transforming how it plans to fight in this new environment. The air component must achieve a requisite level of control in the air to enable BCT success on the ground. Nonetheless, the new context in which air and ground forces operate, accompanied by increasing threats, decreasing budgets, and transformation across DoD, invites new ways of looking at counterair operations.

The Counterair Mission and Culture in the U.S. Air Force

The period between the world wars was formative for U.S. combat aviation. One might say that the U.S. Air Force came of age during the Second World War. It was not, however, yet a separate Service. As Phillip Meilinger wrote, this maturation process under the yoke of the U.S. Army placed high priority on the development of concepts and capabilities that clearly demonstrated independence:

> The primary justification for the Air Force as a separate Service hinged on its demonstrated ability to conduct the independent mission of strategic bombing. Although it could ably support the army, navy and marines—which it had done massively in all theaters throughout the war—it also did something unique. Strategic attack, the bombing of an enemy's vital centers, usually at long range deep inside a defended country, was a capability not possessed by the other Services. Only the Air Force could operate, routinely, at the strategic level of war and produce immediate results. If all the Air Force did was support the army or fleet, then it would seem appropriate for it to remain part of the army and navy (Meilinger 2006, 9).

With an independent Service clearly being the preferred option, the die was cast. Throughout the remainder of American airpower's first century, one could argue that mainstream thought within most airpower circles maintained varying shades of the same initial hue. For the most part, this approach served the nation well in the Twentieth century. The background in the new century's strategic landscape, however, may highlight a contrast between tomorrow's problems and yesterday's approved solution.

When the Air Force became an independent Service at the zenith of strategic bombing doctrine, General Curtis Lemay, his Strategic Air Command, and its bomber force reigned supreme. The organization owning the Air Force fleet of fighter aircraft, Tactical Air Command, which was headquartered at Langley Air Force Base, Virginia, was clearly the junior partner. Through the next few decades, however, the dominant "bomber mafia" began to give way to the "fighter mafia" and air superiority fighter pilots eventually took control of the top Air Force billets (see Worden 1998). The pinnacle of this transition was perhaps the rise of an F-15 pilot, General Tony McPeak, to the Service's top position. The new Air Force Chief of Staff would stand down Strategic Air Command, moving all the Service's bombers and fighters into the same new organization, Air Combat Command, with its home conspicuously located at Langley.

In a speech at Maxwell Air Force Base in June 1992, McPeak would introduce a new Air Force mission as well: "To defend the United States through control and exploitation of air and space." Phil Meilinger recalled the presentation:

> I was sitting in the audience that night and remember feeling a bit deflated at this pronouncement … "Defend the country and gain air superiority; is that it?" I wondered. The Chief tried to explain the meaning behind his simple—and to my mind simplistic—words, but I was no longer listening. Only much later, after reading the actual words spoken that night and then reflecting on them, did the weight of McPeak's speech begin to register. Unfortunately, it is the nature of bumper stickers that they don't come with an explanatory text (Meilinger 2006, 14).

Later in the same manuscript, Meilinger argued that, "in essence, it's almost as if we've substituted the mission of air superiority for the old one of strategic bombing, with little thought given as to how that superiority should be used to achieve national objectives" (Meilinger 2006, 34). To be fair, McPeak's new Air Force mission, featuring both control and exploitation, implied something beyond air superiority, but most airpower advocates read control as air superiority and exploitation as strategic attack. Therefore, air superiority had not necessarily usurped the primacy of strategic attack but had instead joined it at the top of the list. The interwar debate between advocates of Italian airpower theorist Guilio Douhet, who preached the preeminence of control in the air, and the U.S. Air Corps Tactical School, which advocated the strategic bombing of the enemy's vital centers, had come full circle. Still out of the running was the airpower mission of attacking the enemy's fielded forces, which remained a distant third. A central argument in this study is that Army transformation and the age of interdependence bring that prioritization scheme up for negotiation once again.

As it continued to develop later in the Twentieth century, the concept of air superiority became what might be called an article of faith in the U.S. Air Force. Deeply rooted in the theories of airpower that berthed the Air Force as a separate Service, one could argue that the idea of dominating the third dimension had become part and parcel of Air Force culture. Traditionally a central theme in Air Force doctrine, air superiority was until very recently recognized as an Air Force core competency. The transformation of such deeply rooted ideas may require cultural transformation as well. In fact, in order to complete the on-going DoD transformation, transforming Service culture within the Defense Department will be necessary to shift the focus from traditional approaches to a new Twenty-first century paradigm. In the forward to *Rethinking the Principles of War*, retired Navy Vice Admiral Arthur Cebrowski explained that "changing culture is at the heart of transformation more so than any technological or organizational elements" (Cebrowski 2005, xi). Transforming culture lessens the allure of legacy theory, practices, and doctrine, allowing the development of transformational concepts and new ways of fighting in a changed strategic environment. Changing culture, then, will enable the completion of the transformation as a whole.

Air Force doctrine continues to develop and incorporate new ideas. There is a well-established process of merging novel concepts with existing doctrine. This method allows doctrine to grow and to shift in an incremental fashion over time. This approach was largely compatible with changes in the Twentieth century. Evolutionary by its very nature, though, it arguably does not allow for revolutionary thought. As such, continuing to deliberately build upon legacy doctrine may not fully enable transformation. Airpower theorists during the interwar period, such as Douhet and instructors at the Air Corps Tactical School, developed the baseline upon which the U.S. Air Force built airpower doctrine throughout the last century. This doctrine served the nation well in the last century. One could argue that the strategic environment has changed so fundamentally over time, though, that a new framework may be required. In the second century of American airpower, air strategists must lay a new foundation in order to be truly transformational.

Air Superiority from the BCT Perspective

The Air Force *Operations and Organizations* doctrine document draws a line in the sand with regard to air superiority: "From an Airman's perspective, one issue remains preeminent across the range of military operations: air superiority is the desired state before all other combat operations", adding that "operating without air superiority radically increases risk to surface and air operations" (*Operations and Organizations* 2006, 21). Few would argue with either assertion. The real issue is what level of control in the air must be achieved and what level of effort must be committed to that mission. More specifically, the risk associated with air superiority is not calculated in a vacuum. Instead, it must be viewed relative to the risk that may be assumed elsewhere if an exceedingly high level of effort focused on counterair pulls assets away from other priority missions.

This issue of relative risk may be further illuminated by an Army perspective on air superiority. Army Colonel Hymes, for example, explained that "the unpredictable nature of combat operations means that fighting an enemy with an air capability … will result in a loss or losses of U.S. Soldiers. We will not be deterred by that fact" (Hymes 2006). Army Lieutenant Colonel Anderson agreed: "I think folks understand that there's always that chance where one or two guys may get through and may hit us—it's kind of the cost of doing business. It's not a perfect situation where the Air Force is going to get them all" (Anderson 2006). These observations suggest the Army may be willing to assume more risk in the counterair mission than the Air Force has typically accepted in the past. Recent trends toward air supremacy in Air Force circles highlight the point. Army Colonel Pete Curry clearly articulated where there may be differing opinions between the Services on the need for air supremacy: "I think, in the grand scheme of things, air superiority is the goal, the desired end, the 85 percent solution. Air supremacy could take a lot of time and energy. It might matter to you guys but, to us, it doesn't matter" (Curry 2006).

According to Phillip Meilinger, the Air Force should take great pride in its astounding record of controlling the skies over America's wars:

> For over sixty years U.S. forces have enjoyed the luxury of conducting military operations under the umbrella of air superiority. In the past three decades that superiority has extended into space; moreover, superiority has now given way to air and space supremacy—the Air Force has not lost in air-to-air combat since the Vietnam War … An American soldier has not been killed by an enemy aircraft since the Korean War. The U.S. Army has never shot a SAM at an enemy aircraft; none have been allowed to get that close. That is an amazing record (Meilinger 2006, 15).

It also may be a double-edged sword. By some reports, the U.S. Air Force is *184-0* in air-to-air combat since the Vietnam War. Such overwhelmingly decisive odds may invite some to ask about assuming marginal level of risk in the counterair mission and shifting limited resources elsewhere. It is interesting to note that, as part of their on-going transformation, the Army plans to reduce short-range air defense battalions by as much as 60 percent (St-Pierre 2006). While it demonstrates their faith in the Air Force's capacity to deal with the enemy air threat, this move also suggests a willingness on the Army's part to assume increased risk in this area.

The Air Force's profound success in the counterair mission unquestionably demonstrates the Service's commitment to protecting its brethren on the ground. It is also often used to demonstrate the importance of air superiority. The *DoD Dictionary* defines *air superiority* as "that degree of dominance in the air battle of one force over another that permits the conduct of operations by the former and its related land, sea and air forces at a given time and place without prohibitive interference by the opposing force" (*DoD Dictionary* 2001). *Air supremacy*, on the other hand, is "that degree of air superiority wherein the opposing air force is incapable of effective interference" (*DoD Dictionary* 2001). As the Armed Forces look to war in the Twenty-first century, it may be necessary to closely analyze what might constitute "prohibitive interference." It is instructive that, in a survey of field-grade Army officers, 100 percent responded that the death of one U.S. Army Soldier as the direct result of an enemy air attack would not constitute prohibitive interference. St-Pierre articulated what might be considered a representative Army position on prohibitive interference:

> Zero casualties is too much to expect. Losing any soldier is regrettable but people get killed in war. Losing one soldier—even ten soldiers—to enemy air attack is not the issue. As long as we can keep the enemy's air force from preventing us from achieving our tactical, operational, and strategic goals, losing a certain number of soldiers, although quite regrettable, would not be prohibitive. A soldier lost to enemy air attack is no different from our perspective than one lost to sniper fire (St-Pierre 2006).

In fact, over 77 percent of Army officers responding to the survey found the loss of up to ten Soldiers to air attack permissible under most circumstances. According to Army Colonel Marcus De Oliveira, "in a pure attrition based scenario, the loss of up to ten U.S. Army Soldiers as a direct result of enemy air attack would not necessarily constitute 'prohibitive interference.' The loss of a certain ten Soldiers at the decisive place/time, however, could be prohibitive." He went on to define prohibitive in this context as simply "when losses impact the BCT's ability to achieve the commander's intent" (De Oliviera 2006). Hymes submitted that prohibitive interference was "conditionally dependent on the political and operational nature of the event, but I'd have to say [in the event of up to ten casualties] we would replace our losses and continue operations" (Hymes 2006).

Zielinski went so far as to say that, "if enemy air took down thirty percent of a unit but that unit still could accomplish its mission, the losses would not be prohibitive." He then tied air superiority once again to the overall objectives by accepting "a ten percent loss in another unit could be prohibitive if they couldn't achieve the mission" (Zielinski 2006).

Army Colonel Robert Akam explained that, "unfortunately, casualties are part of any armed conflict, and although we have dominated enemy air forces since Vietnam, there may well be a time in the future where potential adversaries have some air capability without having air superiority." In such a scenario, he argued that friendly efforts to ensure zero ground losses from air attack would "give that potential adversary an easy seam to exploit between the military, its doctrine, and our civilian population. I am not sure what a reasonable number of U.S. casualties would be, but back in the days of the Fulda Gap scenarios, these numbers would have most likely been in the hundreds in the early hours and days of conflict" (Akam 2006). Akam's reference to Soviet-era AirLand Battle provides a useful baseline for analyzing acceptable levels of interference in the Twenty-first century version. He went on the postulate that, if five to ten percent of total friendly casualties in mid- to high-level conflict were due to enemy attack from the air, then such interference would be considered prohibitive (Akam 2006). Of the 146 American Servicemembers killed in Desert Storm, then, such a formula would allow as many as 14 deaths due to enemy air attack under an umbrella of air superiority. Airmen might argue this point, but it serves to highlight a fairly substantial difference between Army and Air Force perceptions of allowable risk. More to the point, while these Army officers have clearly indicated their willingness to accept a relevant number of casualties, the Air Force has not allowed a single loss in over 50 years.

The level of risk these Army field-grade officers are willing to assume, then, may allow the Air Force to refine its collective thinking on what could be considered permissible under the title of air superiority. As is currently the case with the F-15E, aircraft such as the F-22 and F-35 Joint Strike Fighter will have the capability to perform both air-to-air and air-to-ground missions. For the most part, the air tasking order (ATO) normally assigns a particular aircraft or formation of aircraft only one of these tasks as the primary mission and directs the associated weapons load accordingly. As such, while these aircraft will have the capacity to protect themselves while dropping bombs, they will be tasked with either the counterair or the counterland mission. As the Air Force continues to reduce the number of total combat aircraft—and a larger percentage can serve in multiple roles—serious consideration should be given to the impact of a high level of effort on the counterair mission. For example, rather than flying combat air patrol missions once air superiority has been secured, a four-ship of F-15Es—and later F-35s—might instead be assigned to an air-to-ground mission. The formation may destroy six enemy tanks during the same time period they would have otherwise kept the skies free of enemy fighters.[1] There is, of course, a slightly increased chance that one American Soldier might be lost to an errant enemy air attack as a result of this decision. Meanwhile, as many as 50-100 friendly lives could be spared by friendly air-to-ground attacks on enemy forces and their equipment. According to the Army's stance on prohibitive interference established in this study, air superiority has been maintained in such a scenario. Enemy ground forces—as threatening as enemy fighters—were destroyed at the same time. Furthermore, such a decision would also likely move the operation closer to achieving theater objectives.

Similar analysis of prohibitive interference as it relates to the loss of friendly aircraft might yield similar conclusions. In Operation Allied Force, for example, two Allied aircraft, an F-117A and F-16CJ, were lost behind enemy lines during 78 days of combat air operations. Due to extraordinary heroism during two combat search and rescue missions, both pilots were recovered and safely returned without serious injury or incident. The loss of military assets was unfortunate and did lead to some changes in procedural guidance. Nonetheless, F-117As and F-16CJs flew almost identical missions over precisely the same target areas for several weeks following both incidents. Simply stated, enemy air defense attacks on Coalition aircraft did not constitute prohibitive interference. The Coalition air operation continued without substantive change or drastically increased concessions to the threat posed by enemy air defenses. In conclusion, then, it is reasonable to consider the loss of a few Soldiers and a few aircraft to enemy air attack acceptable under the umbrella of air superiority. While the hope would be that no friendly lives or equipment would be lost, such is the nature of warfare and the efficient use of limited resources elsewhere would mitigate far greater potential losses in other areas.

BCT Employment and Relative Control of the Air

Joint Publication 1-02 defines *counterair* as "a mission that integrates offensive and defensive operations to attain and maintain a desired degree of air superiority. Counterair missions are designed to destroy or negate enemy aircraft and missiles, both before and after launch" (*DoD Dictionary*). Significantly, this definition links the counterair mission to a desired degree of air superiority, implying the existence of a spectrum of control in the air. Such a conceptualization might have enduring doctrinal utility and would also lay a useful foundation for the associated

---

1 There may be some correlation between this idea and the current practice of "push CAS" where aircraft flying counterair missions such as combat air patrol are made available to conduct counterland missions during the same sortie. However, the point here is that assuming some increased risk in the counterair mission may allow a significant number of multi-role aircraft to focus exclusively on the counterland mission, thereby more efficiently employing these limited resources.

development of *AirLandBattle21* counterair concepts. Unfortunately, Air Force doctrine does not currently establish such a spectrum.[2]

As such, Figure 6.1 offers a novel method of viewing differing degrees of counterair along a linear continuum.[3] The enemy's ability to interfere with friendly air and ground operations is greatest on the far left. There is no enemy interference at the far right. Friendly control of the air, of course, is the reciprocal of enemy interference, with hypothetically no friendly air control on the left side of the spectrum and maximum control to the far right. Reading from left to right, *local air superiority* offers the lowest defined level of relative friendly control in the air, implying a relatively higher level of potential enemy interference. The term *local* as it relates to air superiority in this text implies the achievement of air superiority in a specific area or for a specific period of time. Local air superiority, then, would allow no prohibitive interference in a particular area but would provide the least amount of control across the theater as a whole. The 2003 *Air Force Basic Doctrine*

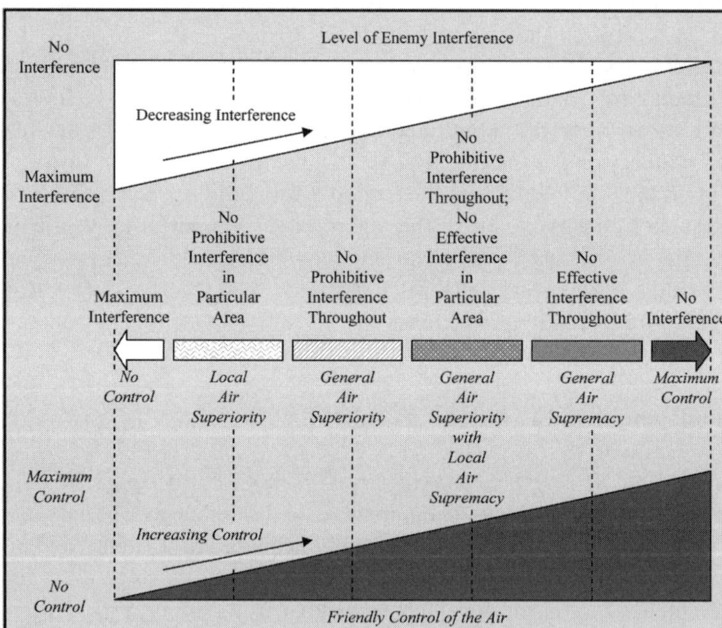

Figure 6.1 A spectrum for relative control of the air

2 As the Air Force endeavors to update current doctrine documents such as the 2002 *Counterair Operations* publication, concepts are beginning to emerge that recognize differing levels of air control and the utility of a more localized option.

3 See, Hinman (2007) *Toward AirLandBattle21: Transformational Concepts for Integrating 21st Century Air and Ground Forces* and Hinman (2008) "Counterair and Counterland Concepts for the 21st Century".

document offered a pertinent discussion of air superiority and supremacy that ends with its only reference to local air superiority: "Various degrees of control are possible ... While air and space supremacy is most desirable, it may exact too high a price. Superiority, even local or mission-specific superiority, may provide sufficient freedom of action to accomplish assigned objectives" (*Air Force Basic Doctrine* 2003, 77). Indeed, air supremacy may unnecessarily tie up limited resources, as the doctrine stated, and local air superiority may be sufficient to achieve the mission. These two critical points are sometimes lost on those who wear Air Force blue. They also lay the foundation for key counterair precepts of *AirLandBattle21*.

Douhet wrote in 1927 that, "in order to assure an adequate national defense, it is necessary—and sufficient—to be in a position in case of war to conquer the command of the air ... Any diversion from this primary purpose is an error" (Douhet 1983, 28). Achieving some level of command of the air is certainly an important objective in war. It is, however, not the only objective as Douhet so emphatically argued. Contrary to his most fundamental precepts, some command of the air is almost always necessary but is almost never sufficient.[4] Later in the same text, Douhet predictably shed a negative light on relative command of the air (Douhet 1983, 98). Parenthetically, some might refute the utility of discussing concepts developed by an Italian airpower theorist who wrote over 80 years ago. Douhet is unquestionably among the founding fathers of U.S. Air Force doctrine. His influence was clearly evident throughout the entire Twentieth century. A new strategic environment radically different from the interwar period in which he wrote offers the visionary thinkers of this new century the opportunity for innovative thought as revolutionary as Douhet but wholly unbounded by his antiquated assertions.

General air superiority would constitute the next level in the spectrum of relative air control. The term *general* in this context connotes constant air superiority across the entire battlespace without any temporal restrictions unless otherwise specified. For example, an air strategy might call for general air superiority throughout the whole theater, with the exception of a particular area that, for some reason, will only require local air superiority during a certain specified time period. While Douhet was advocating absolute command of the air, another interwar theorist, John Slessor, more accurately found that "air superiority is only a means to an end and, unless it is kept in its proper place as such, is liable to lead to waste of effort and dispersion of force" (Slessor 1936, 4). Achieving an unnecessary level of air superiority or one incongruous

4 To be fair, Douhet later makes an important concession that is often missed by his followers: "I have always maintained that the essential purpose of an Air Force is to conquer the command of the air by first wiping out the enemy's air forces. This, then, would seem to be always the first objective of an Independent Air Force. But this is not always the case. Take, for instance, a case when the enemy's aerial forces are so weak it would be a waste of time to devote men and materials to so unimportant an objective. It may be more profitable to subject the enemy to various other offensive actions instead, thereby doing him far more damage" (Douhet 1983, 51).

with the overall military strategy would potentially miss the opportunity to more appropriately mass airpower elsewhere. Slessor fully recognized the elusive nature of such a counterair-centric effort and, in so doing, articulated a position on prohibitive interference much more in line with the collective Army perspective discussed in the previous section:

> Once warfare gets into the third dimension, whether above or below the surface, the cubic area of the battle-fields is so immense that absolute command is hardly ever practicable … the surest method of securing freedom of action for our own aircraft would be the destruction of the enemy's air forces, if that were possible. But as in naval warfare the most that can normally be secured under modern conditions is the neutralization of the enemy's battle fleet, so also in the air. The unlimited opportunities for evasion have the result that the most we can hope for in a war against a first-class enemy is to throw his air forces on the defensive, to neutralize them, to enable our own aircraft to work with the minimum of interference, and to reduce enemy air action against our own forces to the minimum. And even this limited result can only be secured by hard and continuous fighting (Slessor 1936, 5).

In a variation of the concept discussed in the previous chapter, one could argue that an air strike deep into enemy territory exposes friendly airpower's flank. The goal of air superiority is an effort to protect these flanks. The concept of localized air superiority would be analogous to placing less than unequivocal priority on the security of these flanks. Both seem to fit Slessor's general conception while antithetical to Douhet. Because of the inherent speed and flexibility of airpower, these supposed third-dimensional flanks would be quite fleeting and arguably defended through the self-protection of modern fighter aircraft, stealth and other traditional methods such as fighter sweeps and electronic attack. Remaining with the same analogy, general air superiority would be roughly equivalent to a more traditional ground attack on a contiguous front, where air forces would, in essence, maintain the integrity of a vertical front until the air battle is won and air superiority is achieved across the depth and breadth of the battlespace. Localized air superiority, then, is a more audacious approach that requires the commander to assume some risk in the counterair battle. As it is with most bold moves in war, along with increased risk comes the opportunity of increased gains. In this particular case, some risk in the air battle may yield substantial gain in the battle on the ground.

The next increased level of control would allow for *local air supremacy* in a specified location while maintaining general air superiority across the entire theater. As the definitions of air superiority and supremacy suggest, this construct would allow no effective interference in the particular area assigned local air supremacy and no prohibitive interference throughout the theater as a whole. The spectrum of control in the air, featuring this concept of *general air superiority with local air supremacy*, provides a framework for the efficient use of limited air assets, allowing air strategists maximum flexibility to plan for only the necessary levels of air control in different areas across the theater. Naturally, the next level of control, allowing no effective enemy interference throughout the entire battlespace, would be *general air supremacy*. While a fully legitimate goal in some circumstances, air planners all too often aim to achieve this extremely high level of air control before it becomes necessary to do so, inefficiently diverting air assets that may be more effectively used elsewhere.

The concept of local air supremacy may appear to be a departure from generally accepted airpower principles. In fact, it is; transformational thought should generate such ideas. Nonetheless, the notion does have an evidentiary base as far back as the Second World War when the Soviet Air Force used a similar approach on the Eastern Front. In *The Air War, 1939-1945*, R.J. Overy observed that the Soviet military would concentrate a large aerial force at certain points along the front: "This did not necessarily guarantee permanent air supremacy but it would help to give local and temporary supremacy for a particular battle" (Overy 1980, 54). This concept meshed perfectly with the overarching Soviet air strategy focused primarily on targeting German airfields and fielded forces directly associated with a particular battle. Thus, focusing on specific military objectives allowed for the concentration of Soviet aircraft and the achievement of local air supremacy as presented in this study. Of note, Soviet counterair attacks were not an end in themselves but rather intended as a means that enabled the destruction of German ground forces.

Building upon this historical foundation, the widespread application of stealth technology during future conflict will further enable this concept of local air supremacy. The presence of an increased percentage of low-observable fighter aircraft, such as the F-22 and the F-35, with the capacity of offensive self-protection, will improve their ability to achieve local air supremacy in a temporal sense. As such, it will be less likely that enemy air defenses will effectively interfere with air operations in a particular sector involving stealth aircraft. Transformation within the Air Force has ushered in truly watershed technological advances as represented by its Twenty-first century stealth fighters. The counterair concepts discussed in this study aim to provide the new ideas that such drastic change requires.

Recently, the term *air dominance* has gained popularity in Air Force circles. Air Force doctrine submitted an interesting explanation of this emerging concept:

> While air supremacy relates to the air domain, air dominance is the highest level of air control that allows us to focus on affecting *surface* events. Air dominance does not require unopposed use of the air domain, but indicates a level or control in the air that allows us to apportion assets against surface objectives, reducing the ability of enemy forces to effectively employ on land or sea. In other words, air control describes our level of ascendancy in the air domain, while air dominance goes further to describe how we use that level of air control to affect enemy surface forces (*Counterland Operations* 2007, 11-12, emphasis in original).

By defining air dominance as a separate and distinct entity, the concept as currently presented in Air Force doctrine is incompatible with a spectrum of relative control of the air. This study has consequently not addressed it in that context. The relationship in this new definition between air control and surface events is noteworthy and an encouraging step in the right direction. However, the suggestion that "the highest level of control" must be achieved in order to shift focus to attacking the enemy's ground forces may invite closer scrutiny. The counterair strategy in *AirLandBattle21* takes a different approach. It promotes nothing more than an adequate or sufficient level of air control, instead advocating what might more accurately be referred to as the *lowest* necessary level. Meilinger explained that "even air and space dominance should not be seen as an end in itself. Once dominance is achieved, it must be exploited. Numerous air and space roles are essential to the successful accomplishment of a joint military operation, but which, like air superiority, are often taken for granted" (Meilinger 2006, 16). Meilinger went on to list a variety of missions that fall in this underemphasized category, including most notably those associated with direct attacks on the enemy's ground forces.

The widespread use of the term *air dominance*, which has no approved joint definition and is not well understood outside Air Force doctrine circles, arguably gives the impression in the joint community that the Air Force has "upped the ante" with a level of air control beyond even general air supremacy. Webster's defines *domination* as "supremacy or preeminence over another."[5] It follows, then, that air dominance would imply maximum control in the air. Where air superiority allows no *prohibitive* interference and air supremacy allows no *effective* interference, air dominance might seem to some to suggest something akin to no enemy interference whatsoever. While perhaps possible in a theoretical sense, such an excessive level of control would be impractical at a minimum and arguably impossible to achieve in most medium- to large-scale combat operations against a credible enemy air force. In fact, migration toward this new concept might lead some to consider wasting limited airpower resources on this evasive task. Consequently, to allay such confusion and avoid ideas antithetical to *AirLandBattle21*, the term *air dominance* is not used elsewhere in this study. General air supremacy is the highest relative level of air control discussed in the pages that follow.

Building on Figure 2.2, which depicted the likely employment of BCTs in a non-contiguous, geographically separated fashion, Figure 6.2 shows a scenario featuring local air superiority in particular areas. This lowest level of air control suggests a situation where air superiority is appropriate or necessary only over friendly territory and in two particular BCT areas of operation. There may be some compatibility with this concept and joint doctrine *Phase I Deter* operations that might feature the deployment of these two BCTs in order to perhaps establish air points of debarkation (APODs) and prepare for follow-on operations. The additional three AOs, that may be activated later, would likely require local air superiority as well as BCTs moved into those locations.

5 There existed no definition for the terms *air dominance*, *dominance*, or *domination* in the 2001 *DOD Dictionary* as amended through 8 August 2006.

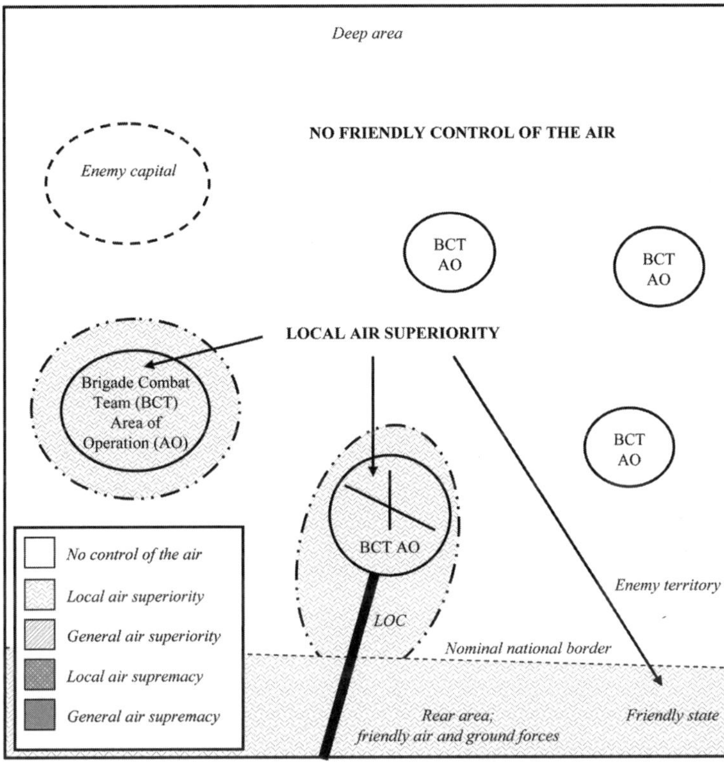

Figure 6.2 Local air superiority

The next level of control in the air would be general air superiority (see Figure 6.3). While such an approach would ensure no prohibitive enemy air interference across the entire theater of combat operations, there may in some circumstances be particular areas that do not initially require this increased level of effort at all times. As Figure 6.3 suggests, examples might include the deep area or a heavily defended enemy capital where constant air superiority would at times be inefficient and unnecessary, particularly in a situation where the bulk of air assets are attacking military targets more directly relevant to friendly surface forces. In such a scenario, local air superiority in a temporal sense would be planned when required to enable air strikes into these areas. While there are multiple conceivable variations to this concept, an air strategy along these lines might be compatible in certain situations with joint doctrine *Phase II Seize the Initiative* operations. In fact, as responses to the survey of Army field-grade officers suggested, general air superiority may be the highest level of control required throughout certain military operations where enabling and accelerating BCT mission success is the primary purpose of the counterair effort.

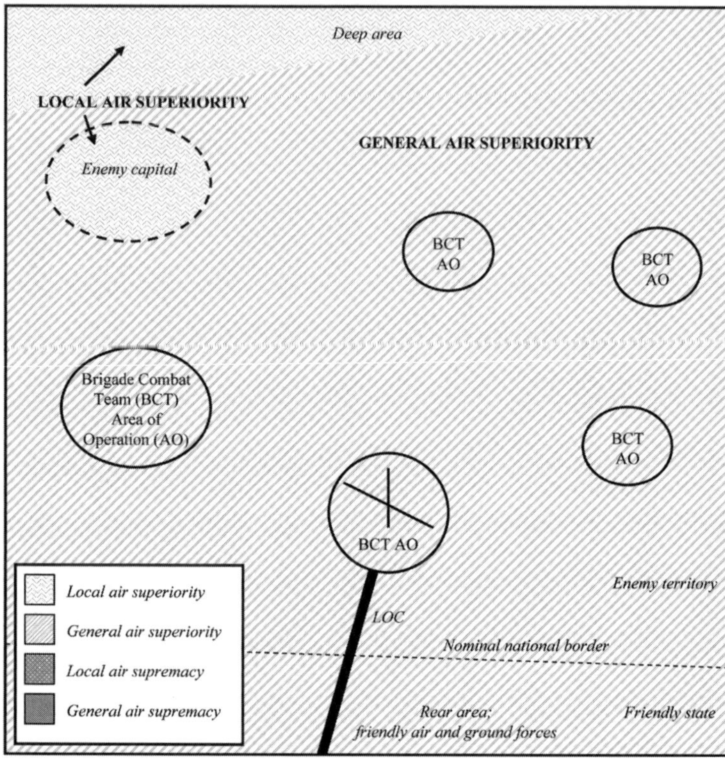

Figure 6.3 General air superiority

In the event that conditions require, the next level of air control, as depicted in Figure 6.4, might feature general air superiority with local air supremacy. In such a scenario, the joint forces commander (JFC) may provide direction that leads the joint forces air component commander (JFACC) to determine that, in addition to general air superiority, local air supremacy is required in certain locations.[6] Examples of areas where no effective interference might be more desirable than no prohibitive interference might include the battlespace over BCT AOs and the rear area. In this increased level of control, it may become necessary to expand general air superiority to areas not formerly offered constant control, such as the enemy's capital and rear area in the previous example. In the event that the situation does not require a higher level of effort, it would be appropriate to remain with the lower level of local air superiority in these areas. The scenario

6 For purposes of uniformity, the general terms JFC and JFACC are used throughout the text rather than Combined Forces Commander (CFC) and Combined Forces Air Component Commander (CFACC).

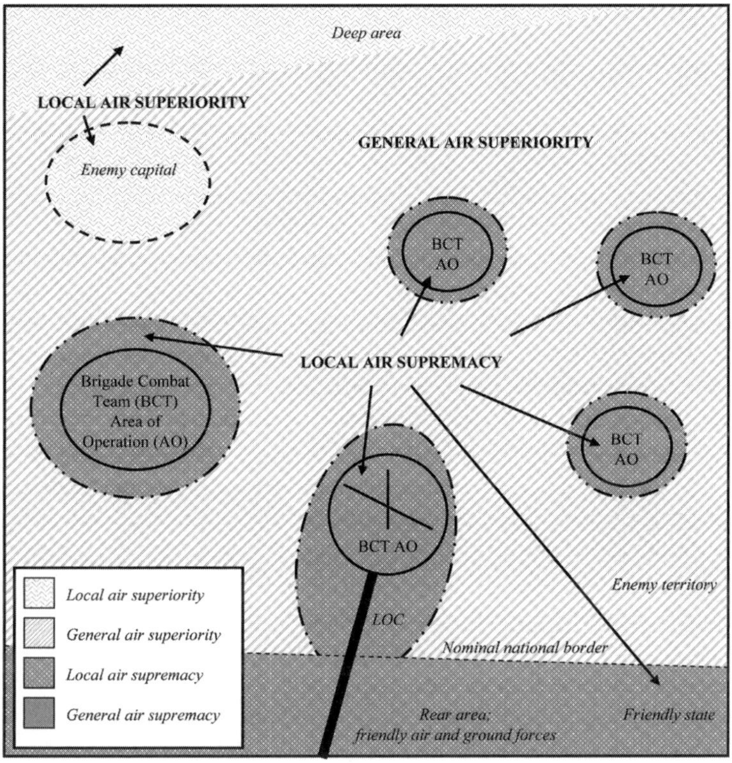

Figure 6.4 General air superiority with local air supremacy

depicted in Figure 6.4 would provide one possible air strategy during *Phase III Dominate* major combat operations, particularly when the JFC's intent placed high priority on joint air-ground operations and focused JFACC efforts on targets of direct military value. During later stages of Phase III, the relevance of targets in the enemy's capital and deep area may drive the need for the expansion of general air superiority into those areas as well. Once again, the construct featuring general air superiority with local air supremacy may be sufficient to address all requirements throughout the remainder of the operation. However, it is conceivable and in some cases likely that *Phase IV Stabilize* operations and *Phase V Enable Civil Authority* efforts would require the absence of effective interference that accompanies general air supremacy (see Figure 6.5).

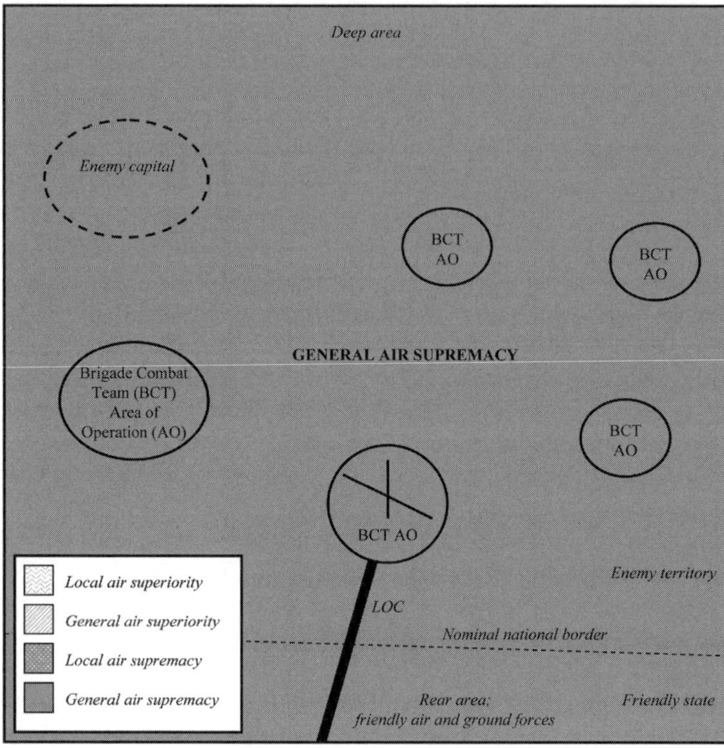

Figure 6.5 General air supremacy

Conclusion

The Air Force requested 381 F-22s to replace an even greater number of aging fighter aircraft in its inventory. Current plans indicate the Service will receive no more than 183 of the new aircraft (Dudney 2006, 2). Although an exceedingly capable new airframe, this drastic reduction in the raw number of fighter aircraft, while demands on Air Force assets are on the rise, must be very seriously considered in planning for future war. Meanwhile, the advent of "double-digit" surface-to-air missiles and sophisticated enemy integrated air defense systems challenges the Air Force's uncontested claim to the third dimension. America's advantage in the air may not be as one-sided as many came to expect in the last century. A basic premise of *AirLandBattle21* is that efforts to maintain this unequivocal advantage in the new century would inefficiently utilize dwindling resources and is generally incompatible with increasingly interdependent air-ground operations.

From the interwar period to the end of the Twentieth century, airpower theory and doctrine placed paramount emphasis on the primacy of the counterair mission. Transformational change in the Army and Air Force, such as the transition to BCTs and to fewer, newer aircraft, necessitates equitable change in this legacy tenet. While airpower advocates promoted the highest levels of air superiority and air supremacy in the past, their Army counterparts now counsel a new approach. Accordingly, general air supremacy may be increasingly less likely and less appropriate in the Twenty-first century. Rather than choosing between the two options of air superiority and air supremacy, future planners using the ideas of *AirLandBattle21* will instead have a scalable menu of options similar to the Army's new modular construct. In order to fight with increasingly interdependent BCTs, Airmen will need to use their limited assets very wisely. Efficiency proved a worthy guide to Airmen in the last century. It will become a necessity in the future.

Chapter 7
Alternatives to Strategic Attack

> In default of knowing how to do what they ought, [they] are very naturally led to do what they know.
>
> Marshall Maurice Comte de Sax

In a 1940 Air Corps Tactical School (ACTS) lecture, Major Muir S. Fairchild developed a fundamental argument common to Airmen of yesterday and today. He found that, in the past, "the ground commander was forced to accept an *intermediate* objective; he was left no choice; he must defeat the enemy's armed force as a preliminary to final military pressure through occupation of the enemy's critical areas." He went on to observe that "this intermediate step to the ultimate goal has been required ever since the beginning of armed conflict. In fact, it has been so consistently necessary as to lead many military men to accept the enemy's armed forces as the true military objective" (Fairchild 1940, 11, emphasis in original). Of course, the military men Fairchild referenced were those who fought wars on the surface. According to Fairchild and fellow ACTS instructors, the advent of airpower gave birth to a new breed of warrior who, from his vantage point in an aircraft well above the ground, looked beyond the enemy's armed force for the true military objective in war. The fact that airpower *could* bypass enemy fielded forces, then, led in the last century to the dogmatic belief in some airpower circles that it therefore *should.*

What began as an ACTS theory during the interwar period arguably persevered in U.S. Air Force doctrine throughout the rest of the century. The beliefs of American airpower's founding fathers like Fairchild berthed similar theories from later prophets such as John Warden. Remnants of the traditional approach and the struggle between old and new remain to this day. For example, the following paragraph from the 2006 Air Force doctrine document on operations is somewhat awkwardly sandwiched between two forward-thinking paragraphs on the AMF concept. Interestingly, all three paragraphs fall under the doctrinal title "Air and Space Maneuver Force":

> In a larger sense, by operating from the third dimension, air and space forces can strike directly at an adversary's centers of gravity, vital centers, and critical vulnerabilities. This capability allows air and space forces to achieve effects well beyond the tactical effects of individual actions, at a tempo that disrupts the adversary's decision cycle. By making effective use of the vertical dimension and time, air and space forces can wrest the initiative, set the terms of battle, establish a dominant tempo of operations, better anticipate the enemy through superior

> observation, and take advantage of tactical and operational opportunities, and thus can strike directly at the adversary's strategy. Air and space power's ability to strike the enemy rapidly and unexpectedly across all of the critical points, from deep to close, adds a significant impact to enemy morale in addition to the physical blow (*Operations and Organizations* 2006, 3).

In *Air Force Basic Doctrine* as well, one could find clear evidence that yesterday's theories of airpower remained prevalent throughout the last century:

> As a concept, strategic attack builds on the idea that it is possible to directly affect an adversary's sources of strength and will to fight without first having to engage and defeat their military forces … Military forces are highly interconnected entities. Through strategic attack, military commanders can directly affect adversary leadership perceptions (either by isolation, deception, or exploitation) and cut off their fielded forces from their leadership and societies, as well as directly attack the adversary's capacity to sustain military forces in the field. While strategic attack may not totally eliminate the need to directly engage the adversary's fielded military forces, it can shape those engagements so they will be fought at the time and place of our choosing under conditions more likely to lead to decisive outcomes with the least risk for friendly forces (*Air Force Basic Doctrine* 2003, 40-41).

This statement makes several arguments central to the U.S. Air Force doctrinal position on strategic attack. It equates strategic attack to attacks on leadership, national will, and war-sustaining capability. Fundamentally, it offers strategic attack as an alternative to attacks on the enemy's military forces. More specifically, it implies that strategic attack will set the conditions for and, therefore, precede any attacks on the enemy's fielded ground forces. Thus, current Air Force doctrine, in line with Fairchild and his ACTS brethren, clearly offers airpower's deep-strike capabilities as an alternative to attacks on ground forces. Instead, it may be more beneficial, advantageous, and appropriate in the Twenty-first century to recognize the potential strategic impact of the same aerial force when integrated with friendly ground forces closer to the battlefield.

Dissecting Decapitation

In the later decades of the last century, airpower enthusiasts began to rally around a concept they called *decapitation*. John Warden popularized the idea with the plan that he designed for the Desert Storm air campaign. He aimed to paralyze and incapacitate the enemy by destroying the maximum number of political leadership, communication, and selected economic targets in the minimum amount of time. Simply stated, he hoped to cut off the enemy's head somewhat literally by killing its leaders and severing their tie to fielded forces. According to Warden, the relentless shock, surprise, and simultaneity of the decapitation approach would bring the enemy to its knees without the need for directly attacking military targets (See Warden 1995).

John Warden arguably made more substantive contributions to strategic thought and intellectual energy in the Air Force than did any other Service member in the later decades of the Twentieth century. The changed strategic context of the new century invites the development of concepts that go beyond the firm foundation he helped to build. Decapitation as a theory, for example, may no longer hold up to a deeper level of scrutiny. First, if the perceived target is the enemy's national and military leaders, it is unlikely that even the most advanced U.S. air and space power capabilities will ever "get them all." It should be noted, for example, that Saddam Hussein and the vast majority of his political and military leaders survived the supposedly relentless decapitation strikes designed by Warden and implemented throughout Desert Storm. Second, in the event such attacks had been more successful in Iraq or another hypothetical scenario, the network of leaders might have been damaged but not destroyed. Other subordinate officers would likely rise up from the ranks to fill the vacancies left by any successful leadership attacks. As the term *decapitation* implies, airpower theorists of the last century tended to espouse concepts that attacked the enemy system from the top down, focusing primarily on leadership and other strategic targets that gave them power. It is not unlikely, then, for such a method to give rise to a "hydra" phenomenon, where cutting off one head leads to the growth of many others. Consequently, it could be said that, where there are troops there will be leaders and a concept built on the single principle of eliminating these leaders welcomes the development of alternative approaches.

For example, an idea more compatible with warfare in the Twenty-first century might turn decapitation on its head, so to speak, and attack the enemy from the bottom up. Such an approach might simultaneously and continuously attack fielded forces and their military leaders at all levels of command from tactical to strategic from the first to the last day of the operation. If possible to truly decapitate enemy leadership, it is arguably more likely to occur in this construct. Of course, the hydra of subordinate leaders may still survive in some weakened form, but the relentless nature of such an attack would paralyze even the most resilient enemy (See Pape 1996).

Targeting the Enemy's Fielded Forces

While ACTS was advocating strategic bombing of enemy vital centers, an approach wholly independent of the fight on the ground, other airpower advocates, such as Royal Air Force officer John Slessor, fully acknowledged the importance of military targets: "The object of the air force in a campaign of the first magnitude in which great armies are engaged is the defeat of the enemy's forces in the field, and primarily of his army" (Slessor 1936, 61). He went on to quite eloquently articulate this position:

> It is absurd to try to make out that air forces can never have any effect in contributing to a decision in a land campaign, except as a sort of super-long-range artillery on the battlefield. It is just as unwise to claim that the decisive point is never "thc Battle" … Consider what these claims amount to: on the one hand that air forces can and must only and always be used for one purpose; on the other, that it is under any circumstances justifiable to split the available air force into two distinct and more or less permanent divisions. Both claims show unmistakably a complete failure to appreciate the meaning or importance of *the principle of concentration*, or to understand the most fundamental and valuable quality of air power—mobility. The answer is, of course, that we must make full use of the mobility and *tactical flexibility of air power*; and *concentrate the maximum force* on whatever task is likely to be decisive, or to contribute most usefully to an ultimate decision, *at the time* (Slessor 1936, 70, emphasis in original).

Slessor's argument suggests that the traditional approach of massing airpower on strategic attack is founded on the shaky assumption that the decisive point invariably lies within the enemy's capital. More accurately, it often lies elsewhere. Accordingly, a basic premise of *AirLandBattle21* is that future planners should look to military targets prior to political and economic ones in their search for an approach that will have the most profound impact on the outcome of the war.

According to Slessor, then, the quintessential target is one of direct military value. Warden, on the contrary, developed a popular theory he labeled "inside-out" warfare that may be viewed as the culmination of Twentieth century airpower thought. He envisioned the enemy system as five rings. The innermost ring in Warden's construct was leadership, encircled by infrastructure and then organic essentials. The fourth ring was the adversary nation's population. The outer-most ring, the enemy's fielded military, encircled and symbolically protected the other interior elements (Warden 1995, 47-51). The traditional approach for ground forces was to attack from the outside in, beginning with the enemy's fielded forces. Warden's inside-out approach, on the contrary, first attacked leadership at the center of the five rings and then moved to infrastructure and organic essentials. Central to his argument was that, because airpower could bypass the enemy's fielded forces, it was no longer necessary to attack them. Skeptics of Warden's approach might be driven to the critique that he aimed at the center of the five rings simply because airpower had the capacity to do so.

The question, however, is no longer which targets airpower can attack but rather which ones it should. Interestingly, while air forces can attack heavily-defended political and economic targets typically located in the enemy's capital, such an approach placed a higher priority on the counterair mission and demanded a much higher level of effort on the related target set. Based on the reasonable assumption that an enemy will place highest priority on protecting these types of targets, Warden's form of strategic attack mandated the highest levels of control in the air. Other approaches, particularly those focused on fielded forces, would generally not require such a substantial level of effort. Overemphasizing the counterair mission, followed by a similarly high priority on the strategic attack that it enabled, came at a predictably high price. With limited resources focused elsewhere in this construct, fewer assets were available for targets of direct military value.

There was a time, of course, when airpower had limited effectiveness against ground forces. That era, however, is long past. In general, theories of the last century were based on two intertwined assumptions. The first was that airpower could have decisive effect on the enemy's political will and economic capacity for war. The second was that airpower had reduced capacity for similar operational and strategic effects while attacking the enemy's fielded forces. As Major General Allen Peck suggested in his forward to the 2006 Air Force doctrine document on counterland operations, contemporary airpower technology brings this dated assumption into question:

> Twenty-first century capabilities regarding precision, information technology, space, intelligence, and command and control increase counterland effectiveness. Modern counterland capabilities can help achieve objectives more efficiently than in the past. This provides commanders a flexible capability that makes other applications of military power more effective and may drive an early end to conflict (*Counterland Operations* 2006, iii).

While these capabilities were very much available during both ground wars in Iraq, Desert Storm and Iraqi Freedom taught painful lessons about the importance of attacking military forces as a matter of course and very high priority. Specifically, a focused effort on the enemy's fielded forces avoids leaving them largely intact as a conventional force, such as what occurred following Desert Storm, or as a guerilla/insurgent force, as with Iraqi Freedom. Decisions made in the closing hours of the "hundred-hour" 1991 ground offensive left the Republican Guard, the Iraqi Army, and Saddam's regime bloodied but still very much intact. Analysts typically cite the Bush administration's decision to cease attacks on Iraqi forces retreating along the "Highway of Death" in Kuwait as a primary reason. On the contrary, higher prioritization of military targets from the first to the last day of the operation would have arguably greatly weakened the Army, Republican Guard, and the regime. Instead, what ensued was twelve costly years of air operations designed to contain Saddam's remaining military power. During this time, he used the military forces that survived Desert Storm to threaten his neighbors and to attack his own people in both northern and southern Iraq.

Ultimately, of course, the U.S. fought yet another ground war in Iraq against many of the same forces that survived Desert Storm. During "round two" in 2003, elements of the Iraqi Army survived once again, this time by throwing off their uniforms and blending into the civilian population. Many were members of Iraqi units bypassed by U.S. Army brigades racing to Baghdad. Thousands survived OIF major combat operations, later opposing U.S. forces in an unconventional manner as terrorists, insurgents, and members of militia groups. It is widely held that many former Republican Guard officers participated in and, in many cases, led aspects of the ensuing insurgency. These combatants proved to be far more formidable as insurgents than when they constituted a regular force. Arguably, a substantially higher level of effort against military targets and bypassed forces during the war's initial phase would have lessened the likelihood of a viable insurgency and weakened the potency of such a movement had it occurred. Additionally, these Iraqi combatants have underscored the fact that airpower is generally far more effective against conventional fielded forces fighting major combat operations than it is once the same people opt for irregular war as insurgents.

Neither situation demonstrated during Desert Storm and OIF is unique to Iraq. Among the most critical lessons from operations in Iraq applicable elsewhere is the realization that the enemy's military forces and equipment must suffer appreciable levels of attrition. When the opportunity presents itself in the battlespace of the future, then, modern air and ground forces must place high priority on defeating the enemy's Army. In fact, with the advent of the distributed operations model, it is even more likely that large numbers of enemy forces will be bypassed in the non-linear, non-contiguous battlespace of the future. As such, conscious and deliberate effort on the part of air and ground forces will likely be necessary to avoid a repeat of the lessons from the wars in Iraq.

Ever since airpower's capabilities matured during the period between the two world wars, practitioners and theorists alike have argued over the importance of three airpower missions referred to in today's Air Force doctrine as counterair, strategic attack, and counterland. As previously discussed, strategic attack and counterair struggled for primacy but counterland remained a distant third. Army Colonel Peter Zielinski captured the essence of this struggle: "I don't know why the Service doesn't emphasize both strategic attack and the mission of attacking enemy ground forces. The Air Force does the close air support mission exceedingly well but the other missions like strategic attack and air superiority seem to get all the attention from a resource perspective and otherwise" (Zielinski 2006). *AirLandBattle21* recommends a new approach that emphasizes counterland as a top priority from the outset of hostilities to the termination of major combat operations. Recognizing resource constraints, this novel framework calls for only the minimum necessary level of control in the air required for each phase of the conflict. Additionally, under this new construct, traditional approaches to strategic attack would be added to the mix when necessary and appropriate. Of course, certain future scenarios may necessitate some level of strategic attack earlier in the operation. However, a phased approach to strategic attack such as the one advocated here may allow a reduced initial level of effort in the counterair mission. If friendly air forces are not attacking a heavily defended enemy capital around the clock, for example, a reduced level of air superiority in this region and elsewhere may be acceptable to the JFACC and joint force commander. Applying an economy-of-force approach to both strategic attack and the counterair mission in turn allows an increased commitment to the counterland mission.

Toward a Transformational Alternative to Strategic Attack

During the Second World War, limitations in technology and doctrine reduced the heavy bomber's efficacy against the enemy's fielded forces. Allied air strategists during the interwar period held that bombers were more effectively and efficiently employed against strategic targets than they were attacking deployed troops, equipment, and supplies. Smaller attack aircraft were more capable against these military targets but the capabilities of the day were viewed as limiting their impact as well. While accurate to a certain degree during the Second World War, these suppositions are by no means true today. With its suite of sensors, data link, and precision weaponry, Twenty-first century American airpower is exceedingly potent against fielded military forces. Nonetheless, prevailing thought in most airpower circles today remains wedded to Twentieth century concepts founded on the obsolescent assumption that this target set is elusive to airpower.

Similarly, interwar theorists generally assumed that the limited capabilities of air forces at the time kept airpower's impact on fielded military forces at the tactical level. Along the same logic, the air arm's only chance at substantial operational or strategic effect was a markedly different approach, such as the industrial web theory, which attacked the enemy's economic infrastructure. Once again, with a few notable exceptions, one could argue that such beliefs were often generally accurate during the Second World War and in the decades that followed. As technology continued to improve, however, airpower's potential capacity for broadened effect increased as well. As is often the case, new technology invited new ideas, and the Cold War doctrine of *AirLand Battle* answered the call. It revolutionized airpower in a sense by moving beyond previous technical and conceptual limitations, thereby enabling aerial targeting of enemy military capabilities, such as Soviet follow-on ground forces, to have aggregate impact at the operational level of war.

In the same manner, *AirLandBattle21* envisions air attacks on targets of direct military value in a way that can now have strategic effect. In fact, the greatly increased capabilities of Twenty-first century airpower have now enabled impact across all three levels of war. The *Air Force Glossary* defines *strategic attack* as "offensive action conducted by command authorities aimed at generating effects that most directly achieve our national security objectives by affecting an adversary's leadership, conflict-sustaining resources, and/or strategy" (Air Force Glossary 2003).[1] Once could argue quite convincingly that attacking targets of direct military value at the tactical, operational, and strategic level would affect each of these areas. In the Twenty-first century, then, attacking military forces can fit the definition of strategic attack when enabled by technological capabilities and transformational concepts that allow it to span the levels of war.

As discussed, Douhet's command of the air and the ACTS strategic bombing formula are at the foundation of U.S. Air Force doctrine. John Warden built his

1 The *Air Force Glossary* offered this definition, from the 2002 Air Force Doctrine Symposium, as the most current and accepted definition of the term.

concepts upon this base. In fact, with the exception of *AirLand Battle* doctrine, his approach could be viewed as the synthesis and culmination of the body of airpower theory widely accepted throughout the Twentieth century. Because his concepts essentially took the same form as those of his predecessors, it could be argued that his ideas, while somewhat novel, were not transformational. Rooted in an outdated strategic context and arguably foreign to the Army's new distributed-operations model, many of the existing theories of airpower fall short of the kind of revolutionary thought necessary to complete the current Defense Department transformation.

In addition to decapitation, John Warden espoused and popularized the concept of parallel war. This idea held that simultaneously attacking targets at the tactical, operational, and strategic levels of war would cause strategic paralysis of the enemy system. Instead of disparate target sets with divergent goals, a strategy following the concept of parallel war would ensure tactical, operational, and strategic attacks were neatly aligned with each of the other levels. Interestingly, though, Warden's approach placed considerably less emphasis on enemy fielded forces. Chairman of the Joint Chiefs of Staff, General Colin Powell, for example, expressed concern over the lack of reference to attacking enemy tanks in the "Instant Thunder" air campaign Warden and his Checkmate planners designed for Operation Desert Storm (Reynolds 1995, 72-74). Instead, Warden opted for political and economic targets of perceived value at the national level. As such, his approach ironically led to a wide array of strategic targets with no parallel at the tactical level. In fact, Central Command JFACC, General Charles A. ("Chuck") Horner also strongly objected to the absence of attacks on fielded forces when Warden later briefed the plan at Air Component headquarters in Riyadh, Saudi Arabia (Reynolds 1995, 129-130).

Furthermore, for parallel war to be most effective, attacks at the tactical, operational, and strategic level would be fully integrated against the same general target set. In most cases, the only targets that would naturally reside at each of these levels are those of a direct military nature. Accordingly, both John Warden and John Boyd argued that an air operation is most likely to achieve strategic paralysis when the plan masses effects against targets directly related to each other and within the same system. In line with their teachings, 2006 Air Force doctrine submitted that "air and space power is usually employed to greatest effect in parallel, asymmetric operations", explaining that "parallel operations are those that apply pressure at many points across a system in a short period of time in order to cause maximum shock and dislocation effects across that system" (*Operations and Organizations* 2007, 10). The enemy system most likely to present targets across the levels of war, and therefore potentially more vulnerable to paralysis, would be the enemy's military. Consequently, the most efficient and effective method of achieving strategic effects in an economy-of-force operation would be massing air attacks against targets most closely tied to the enemy's military capabilities. Traditional approaches to strategic attack, however, often looked elsewhere in hopes of finding that one, single target or neatly-interrelated group of targets that would theoretically end the conflict expeditiously. In the event these attacks did not achieve the desired effect, such a strategy had limited impact on the adversary's military strength or the war on the ground. On the other hand, constant attacks on military targets across the depth and breadth of the battlespace would significantly degrade the enemy military capacity to continue fighting and, by crushing the enemy system over time, could not help but have cumulative effect at the operational and strategic level. Furthermore, airpower's ability to bring massive and focused attrition upon the enemy's ground forces may affect the enemy system as a whole. Accordingly, Marine Corps doctrine asserted that, "at the critical point, where strength has been focused against enemy vulnerability, attrition may be extreme and may involve the outright annihilation of enemy elements. Nonetheless, the object of such local attrition is not merely to contribute incrementally to the overall wearing down of the entire enemy force, but to eliminate a key element which incapacitates the enemy systemically" (*Warfighting* 1997, 38). As such, Boyd's approach is as applicable to the adversary's armed forces as it would be to any other system.

Warden's attempt to serve the two masters of decapitation and parallel war, then, could be viewed as somewhat counterintuitive. While the concept of parallel war hinged fundamentally on debilitating attacks at all three levels of war, Warden-style decapitation focused predominantly at the strategic level and did not fully embrace the viability of attacking tactical military targets. Furthermore, Warden offered few alternatives in the event that decapitation proved unsuccessful. In the approach offered in this study, repeated attacks on fielded forces would, at the very least, have an important cumulative effect in their own right, bringing the enemy's military closer to defeat.

In its 2006 *Counterland Operations* publication, the Air Force acknowledged the very real effects that attacks on the enemy's armed forces can have across these three levels of war:

> Counterland operations can achieve tactical-, operational-, or strategic-level effects, and can significantly influence the course of a military operation. Counterland effects focus at the tactical and operational levels of war by targeting fielded enemy surface forces and their supporting infrastructure. When planned and executed successfully, counterland operational effects will indirectly lead to strategic effects by denying the enemy's ability to execute a coherent ground campaign. In cases where the enemy places strategic value on a specific portion of their ground combat force, counterland operations can produce more immediate effects at the strategic level (*Counterland Operations* 2006, 4-5).

In making this critical point, Air Force doctrine demonstrated the limited ability of current doctrinal terms to holistically address attacks on targets of a direct military nature. The counterland mission, for example, is doctrinally distinct from strategic attack. By definition, the target sets are distinct as well. While CAS is normally associated with the tactical level, air interdiction as currently defined may have operational effects. The strategic realm, however, is, both literally and in current practice, generally reserved for strategic attack. And yet, counterland doctrine clearly acknowledged the real potential for strategic effect by attacking military targets.

Traditional arguments advocating top-down strategic attack submitted that attacks expressly aimed at the strategic level would "trickle down" to the operational level with eventual impact at the tactical level as well. An alternate bottom-up approach, that directly attacks tactical and operational military targets, would instead have immediate, tangible effects at those levels. It may be more likely that the accumulation of these effects would have substantial effect at the strategic level than the other way around. In *Bombing to Win: Air Power and Coercion in War*, airpower theorist Robert Pape advocated such a methodology. The ability of modern airpower to mass effects across the battlespace in a condensed period of time would arguably allow such impact to take hold far more expeditiously than in the past. In the unlikely situation that the impact did not migrate to the strategic level in such a timeframe, an approach along these lines would nonetheless achieve appreciable and lasting effects on the military targets themselves. On the contrary, if traditional strategic attack methods were unsuccessful, there would be no residual impact on the enemy's fielded forces. In such a scenario, then, these aerial efforts would have minimal impact on the ground battle. This situation would be even more unacceptable in today's era of interdependence than it may have been in the past.

Air Force doctrine offered a position exceedingly germane to this discussion:

> Parallel operations should be conducted in conjunction with other elements of a joint force. For example, counterland operations, in conjunction with simultaneous and coordinated attack by surface forces, can overwhelm an enemy's reinforcement and resupply capacity or his ability to command his forces, creating synergistic effects that have an adverse impact throughout the enemy system. In this case, the surface and air and space maneuver elements of the joint force are integrated with each other, rather than one in support of another, to achieve decisive results (*Operations and Organizations* 2007, 11).

Warden's approach to strategic attack, which employed airpower independent of ground forces, would likely attack one set of targets or one enemy system while ground forces attacked another. Rather than parallel and synergistic, such attacks would be disparate and largely unrelated. The interdependent application of integrated air and ground forces against the same targets of direct military value, on the other hand, would maximize the likelihood of a synergistic effect. The air effort would be parallel across the levels of war and also parallel with ground attacks on the same enemy military system. *Air Force Basic Doctrine* explained in 2003 that "commanders can build on the tactical effects of close air support by orchestrating it with other surface and air operations to produce operational-level effects" (*Air Force Basic Doctrine* 2003, 45). As the term synergy implies, the cumulative effects of CAS at the tactical level, BAO at the operational level, and additional attacks on military targets with strategic impact would be far greater than the sum of their individual parts. Operating in conjunction with BCT attacks on similar targets, this combined-arms approach would pack a potent punch unmatched by previous approaches to parallel war or traditional methods of strategic attack.

Counterforce Parallel Attack

The approved joint definition for the term counterforce is "the employment of strategic air and missile forces in an effort to destroy, or render impotent, selected military capabilities of an enemy force under any of the circumstances by which hostilities may be initiated" (*DoD Dictionary* 2001). While there is no current joint definition for the term *countervalue*, the *Air Force Glossary* defines it as "the employment of strategic air and missile forces in an effort to destroy, or render impotent, selected industrial and economic capabilities of an enemy force under any of the circumstances by which hostilities may be initiated" (*Air Force Glossary* 2003). As the definitions imply, these terms traditionally reference the two prevailing theories of nuclear targeting. The countervalue approach, for example, attacks cities with nuclear weapons while the counterforce method would target the enemy's armed forces. Applying the same terms to conventional air attack would add clarity to the previous discussion. Under this construct, Warden's approach, which attacked political, economic, and industrial targets, might be dubbed *countervalue parallel attack*. *AirLandBattle21*, on the other hand, espouses a concept that would perhaps most appropriately by labeled *counterforce parallel attack*. This transformational alternative to traditional approaches to strategic attack would predominantly employ airpower against the enemy's military might at the tactical, operational, and strategic levels of war across the breadth and depth of the battlespace.

One example of counterforce parallel attack would be air operations against the enemy's transportation system. At the tactical level, airpower might interdict tactical mobility, attack military vehicles supporting front-line enemy forces, and target chokepoints and lines of communication (LOCs) relevant to units currently in contact with friendly ground forces. At the operational level, the air component might target bridges on main LOCs supporting thoroughfares that link supply depots and operational reserve forces to projected areas of engagement. At the strategic level, strategists may aim to disrupt what might be called the "strategic movement" of troops, supplies, and equipment throughout the country by systematically attacking the national-level transportation system represented by targets such as train stations and rail yards. They might also target road and rail bridges that link strategic reserve forces or other elite enemy forces to areas where they may be needed or hope to move.

Of course, airpower has attacked transportation targets during many armed conflicts throughout history. The difference here is two-fold. First, the emphasis at the tactical, operational, and strategic level is on targets that directly affect the enemy's military capability rather than a strategically focused economic or political goal with some possible eventual effect at the front lines. Second, attacking such targets within the same system, the transportation system in this example, would cause the kind of paralysis Warden and Boyd envisioned. Paralysis in this case, however, would be far more likely to occur as the attacks are more focused on closely-related military targets with more tangible results and predictable impact. Traditional approaches that may have led to similar targets in the past have generally aimed at strategic effects in their own right. As past attempts have shown, however, paralyzing an entire nation's economy, for example, is more difficult than taking the legs out from under the enemy's military transportation system.

In the Second World War, tactical airpower often focused efforts on the enemy's "first-echelon" front-line fielded forces and achieved substantial impact at the tactical level. AirLand Battle concepts for fighting in the Fulda Gap planned to target second-echelon enemy ground forces with operational-level impact. *AirLandBattle21* aims at attacking both these echelons but adds what might be called "third-echelon" forces such as the strategic reserve and other military targets of strategic effect with the goal of achieving parallel impact at the tactical, operational, and strategic levels. Enemy leaders and their personal elite forces, for example, would constitute strategic, third-echelon military targets. Adolph Hitler's Panzer Lehr of the Second World War and Saddam Hussein's Special Republican Guard of Desert Storm would constitute two examples of elite units that may have received increased levels of early attention had war planners in these past wars applied the strategic principles being offered here.

Once again, recognizing the utility of attacking such targets is not new. These types of targets have been attacked in previous military conflicts with varying levels of emphasis and success. What may be new in counterforce parallel attack is the increased priority of the enemy's fielded forces, the concept of attacking the enemy's armed forces in a parallel nature, and the synergy achieved by air and ground forces massing effects against the same enemy system. Also new is the convenient merger between the air-maneuver force and the concept of counterforce parallel attack. Air and ground forces not only attack parallel target sets but they greatly increase each other's effectiveness when they serve in mutually-supporting roles designed to elicit responses from enemy forces which make them vulnerable to attack.

Traditional approaches to strategic attack may still be applicable in the wars of the Twenty-first century. Indeed, there are future potential scenarios that lend themselves to approaches based on the ideas of Warden and others. The concepts presented in this chapter are not meant to imply otherwise. A transformational concept such as counterforce parallel attack would be more appropriately viewed as widening the body of thought in this area by introducing the potential strategic impact of attacking enemy ground forces in a comprehensive manner. In many cases, commanders and their staffs may opt for this new strategy at the outset of hostilities and consider more traditional forms of strategic attack, such as attacks on political and economic targets, later in the conflict. In other cases, where air resources are available in sufficient quantities, the two may supplement each other from the beginning of the conflict. And finally, this new approach may be applied in place of the old when appropriate to do so, particularly in scenarios featuring limited airpower assets and the increased dependence of the ground component on those forces. The keen interdependence inherent in warfare in the Twenty-first century requires the air component to analyze scarce resources very closely before allocating substantial numbers to traditional methods of strategic attack, especially during the initial stages of conflict.

Prior to the Allied landing at Normandy on 7 June 1944, Hitler positioned elite divisions of the 12 SS and Panzer Lehr as a strategic reserve to front line forces defending the coastline. The Allies could not break out of the beachhead until massive carpet bombing of German fielded forces and attacks against enemy tanks on 24 and 25 July finally broke the back of these forces. Earlier relentless attacks against these forces prior to and immediately after the landing at Normandy would have arguably allowed the Allies to break out much earlier than 27 July. Instead, polarized emphasis on direct support, on the one hand, and independent strategic bombing, on the other, may have delayed Allied efforts to advance beyond the beaches where they were trapped for well over a month.

Ideas that may have been useful in last century's wars seem even more applicable in the future. It is plausible that the models currently associated with strategic attack will not neatly fit the kind of warfare most likely in the Twenty-first century. Many future wars simply may not lend themselves to a wide array of strategic targets of the traditional sort. Instead of a large city replete with political, economic, industrial, and leadership targets, many strategic targets in tomorrow's wars may be more fleeting in nature or hidden away in remote mountain ranges. As such, it may be that the most plausible, practical and visible strategic effects will be achieved by airpower attacking fielded forces compelled into the open by skillfully positioned brigade combat teams.

Chapter 8
From Strategic Attack to Strategic Cordon

> We can't solve problems by using the same kind of thinking we used when we created them.
>
> —Albert Einstein

In addition to counterforce parallel attack, the changed strategic landscape of the Twenty-first century invites other transformational concepts that go beyond traditional methods of strategic attack. *AirLandBattle21* advocates one such idea labeled "strategic cordon." This new approach is designed to provide a range of options for the JFC to seal off—or "cordon"—large areas on the surface such as national borders and geographic regions. By way of explanation, there are similarities between an aerial strategic cordon on land and a naval blockade at sea. While this study recommends BAO as a new Air Force mission along with CAS, air interdiction and strategic attack, strategic cordon and counterforce parallel attack should be viewed as operational concepts that complement these pre-existing missions or as innovative methods of achieving them. As with counterforce parallel attack, strategic cordon is arguably far more compatible with the changed operational environment likely in Twenty-first century warfare than were methods espoused in the past. Rather than attacking strategic targets in the enemy's capital, this new approach will have strategic effect by attacking targets of a more military nature that threaten dispersed BCTs conducting distributed operations across the battlespace. It will allow the JFC to "draw a line in the sand" with airpower, severing avenues of approach between friendly ground forces and enemy troops, such as bypassed forces, who aim to attack smaller, lighter, and geographically separated U.S. Army BCTs. In this way, strategic cordon is not only an alternative to strategic attack and air interdiction but, in a sense, has brought this study full circle with its close tie to the integral *AirLandBattle21* concept of BAO.

In a brief chapter in *On War* dedicated to discussing what he labeled "cordon" operations, Clausewitz offered a few observations quite pertinent to this new concept. First, he made it clear that a cordon would not be impenetrable and could not be expected to stop a large enemy force. It was not designed to repel every enemy attack, and he explained that some commanders in his time had misunderstood its utility: "We must add that there seem to have been occasions when commanders and their general staffs have overlooked the real significance of a cordon-system, treating its relative value as absolute, and believing it capable of holding off all enemy attacks whatever. These were not misapplications of a method, but a total inability to grasp its nature" (Clausewitz 1976, 454-455). Most importantly, though, Clausewitz established the potentially strategic impact of cordon operations. He identified their objective as "the preservation and security of the country against an enemy who is not intent on major action" (Clausewitz 1976, 454-455). While perhaps traditionally considered a tactical effort, stakes this high would arguably elevate certain cordon operations to the level of strategic in effect if not also in application.

Applying airpower to the cordon mission would further enhance interdependence between air and ground forces. In fact, Johnson found real utility in a concept such as strategic cordon as it relates to BCTs and an AMF: "Arguably, an air force could be employed to seal a border … where cover and concealment are very limited and the ROE [rules of engagement] allow sealing rather than mere interdiction. There are lots of factors that would have to go into the execution of such an operation but theoretically it isn't beyond imagination." This method of employing airpower where ground forces may have otherwise been committed will assist in enabling continued high operational tempo and avoiding premature culmination: "With the border area sealed by an air-maneuver force, the … BCTs could be elsewhere employed where their capabilities better mesh with the conditions" (Johnson 2006).

This concept of strategic cordon provides a practical means of addressing the potential prevalence of bypassed forces on battlefields of the future. The new Army model of vertical maneuver, which may catapult BCTs past enemy forces to directly attack other objectives, greatly increases the likelihood of such a problem. Strategic cordon may serve as a viable solution. Bypassed forces, which are likely in future medium- to large-scale combat operations, were addressed in detail in a previous chapter. Consequently, strategic cordon will be discussed in this section as it relates to another challenge—the porous border—which is equally likely across the spectrum of conflict.

Porous Borders in Iraq and Afghanistan

Historically, American airpower doctrine has focused on establishing air superiority, conducting strategic attack, and performing the counterland mission. As discussed previously, each of these mission areas remains important in the Twenty-first century. However, success in the transformed battlespace will be problematic despite these efforts if BCTs engaged in future conflicts are vulnerable to attack in multiple directions from adversaries transiting porous borders. While this study focuses primarily on medium- to large-scale combat operations, features of the war in Iraq may very well find their way onto the conventional battlefields of the future. The 2006 *Iraq Study Group (ISG) Report* found that, in Iraq, "the U.S. military rarely engages in large-scale combat operations. Instead, counterinsurgency efforts focus on a strategy of 'clear, hold, and build'—'clearing' areas of insurgents and death squads, 'holding' those areas with Iraqi security forces, and 'building' areas with quick-impact reconstruction projects" (Baker and Hamilton 2006, 7). This strategy requires that friendly forces pacify the area, provide lasting safety and security, establish infrastructure, and secure the sustained flow of food and medical aid. A safe environment is paramount to allow reconstruction, new police training, and the establishment of local government. In general, the longer that security can be maintained, the more established the local government becomes and the more local support is generated. Even a relatively small number of adversaries infiltrating these areas can bring about substantial setbacks to these efforts.

Traditionally, there have been very few restrictions along Iraq's borders with Iran and Syria. For example, millions of Iranians travel to Iraq each year to visit Shia holy sites. Similarly, the nomadic Shia tribal groups in western Iraq and eastern Syria have traveled frequently and largely unabated between the two countries for generations. Generally unthreatening in times of peace, these porous borders posed a substantial challenge to Coalition efforts in the conflict. As such, the *ISG Report* states that "Syria should control its border with Iraq to stem the flow of funding, insurgents, and terrorists in and out of Iraq" (Baker and Hamilton 2006, xv). Beyond this Syrian influence, the report goes on to highlight the problems associated with Iran and its involvement in Iraq:

> Of all the neighbors, Iran has the most leverage in Iraq. Iran has long-standing ties to many Iraqi Shia politicians, many of whom were exiled to Iran during the Saddam Hussein regime. Iran has provided arms, financial support, and training for Shiite militias within Iraq, as well as political support for Shia parties. There are also reports that Iran has supplied upgraded and improvised explosive devices to groups-including Sunni Arab insurgents that attack U.S. forces. These explosives have proven capable of penetrating the best armored vehicles in the U.S Army. In addition, Abu Musab al-Zarqawi—the leader of al Qaeda in Iraq—has imported foreign fighters to launch suicide attacks in Baghdad. These horrific attacks have killed hundreds of innocent Iraqi men, women, and children as well as members of the Iraqi security forces and Coalition Service members (Garamone 2006, 1).

Sanctuaries in neighboring states provide a safe haven for training and organizing tasks that would not likely be possible in the actual combat zone. Porous borders allow the flow of forces and supplies across lines of communication, often permitting these militant organizations to endure attacks that would otherwise be far more effective (Staniland 2005-2006, 31).

Challenges associated with uncontrolled borders were not unique to Iraq. Coalition forces in Afghanistan confronted the same problems with Taliban and al Qaeda forces fighting from havens in the federally administered tribal areas of Pakistan. More generally, though, this issue of porous borders is likely to impact future operations across the spectrum of military conflict. If the U.S. military is to remain fully viable in the changed strategic landscape of the Twenty-first century, then, it must find efficient and effective methods of dealing with this issue.

The strategic cordon may serve as one such approach. A reliable strategic-cordon capability can limit the effectiveness of the protracted warfare strategies of the enemy. It could lead to fewer tragic headlines detailing the employment of new advanced enemy weapons supplied by outside groups and nations. Strategic cordon may limit foreign-fighter infiltration and reduce the frequency of high-value individuals escaping across borders. Finally, strategic cordon may assist in addressing emerging transnational threats as well as traditional problems of third-nation involvement. As such, the air component can best leverage its assets and unique capabilities to speed the pace of victory throughout the range of military operations by developing the next-generation capability of strategic cordon.

Consequences of Historic Border Failures

Throughout the current conflict in Iraq, Coalition forces have proven largely unable or unwilling to adequately contain the Iraqi border. In fact, American ground forces have been unable to curb transnational freedom of movement across borders in both Iraq and Afghanistan. The recent inability to neutralize this emerging threat reflects a more extensive problem. Of the many negative consequences of uncontained borders, the most serious is perhaps the fact that cross-border attacks, aid, and sanctuary have ceded the advantage of time to the enemy. Past experience in foreign wars has at times featured a lack of American patience manifested by an unwillingness to accept long-term involvement. Cross-border enemy activity generally prolongs the same conflict, bringing America's sometimes fitful commitment into sharper focus. When offensive actions become protracted, resolve at home is often weakened, the enemy becomes emboldened, and politicians move toward short-term resolution rather than long-term change. From the military perspective, the transformed Army aims to drastically accelerate the tempo of warfare in the Twenty-first century while porous borders allow the enemy to play by a different set of rules on their own time table. As in Iraq and Afghanistan, swift victory in future conflicts may be equally elusive if enemy forces are allowed to find safe haven in mountainous regions, seek refuge in urban areas, or escape across national borders to fight again another day. The unchecked flow of new combatants, weapons, and other supplies will further complicate the situation in tomorrow's wars much as they have today.

This problem, however, is by no means simply a Twenty-first century phenomenon. Experience in the past century repeatedly demonstrated the substantive impact of porous borders on military operations. Examples include the influx of Chinese fighters and supplies onto the Korean Peninsula during the Korean War as well as the resupply of North Vietnamese forces from China and through Laos and Cambodia via the Ho Chi Minh trail. The problem is also not unique to the American military. Israeli military officials, for example, found that Hezbollah replenished nearly half of its pre-war stockpiles of short-range missiles and small arms in a few brief months after the 2006 conflict between Israel and

Hezbollah had ended (Shannon and McGirk 2006, 1). In fact, Elaine Shannon reported that "western diplomats say that Hezbollah now has over 20,000 short-range missiles—a similar amount to what they had at the start of the conflict, during which the group is believed to have fired over 3,000 rockets at Israel" (Shannon and McGirk 2006, 1).

The inability to seal borders in combat has significantly contributed to U.S. difficulties in achieving success within timelines favorable to the American people. Retired Lieutenant General David Barno, former commander of U.S. and Coalition forces in Afghanistan, shared the most recent version of a popular saying: "In Afghanistan, Americans have all the wrist watches, but Afghans have all the time" (Scales 2006, 3). Retired Major General Robert Scales, former Army War College Commandant, described the issue of prolonged conflict by adding that "the enemy will attempt to control the clock with the strategic intent of winning by not losing. He will use the clock to wear down American resolve. Management of the clock will allow him to use patience as a means to offset American superiority in killing power. His hope is to leverage our impatience to cause us to overreact with inappropriate use of physical violence" (Scales 2006, 3). In the midst of this battle over the clock, military victory is not likely to be achieved in any reasonable timeframe if the enemy is free to build up forces in foreign sanctuaries, recruit abroad, and funnel combat resources across porous borders.

Traditional and Transformational Counters to Porous Borders

Historically, there have been examples of positive initiatives in this area. Generally, such success accompanied the sustained commitment of national resources and the focused attention from senior military commanders. In the late 1950s, for example, France aimed to curb cross-border support to the insurgency in Algeria with the construction of the Morice Line, a multifaceted physical barrier which spanned the 250 kilometer border between Tunisia and Algeria. The Morice Line consisted of an electric fence that stood eight feet high and was powered by five thousand volts. Barbed wire entanglements and anti-personnel mines covered 50 yards of ground on either side of the fence. A 2006 RAND report estimated that the Morice Line reduced infiltration into Algeria from Tunisia by as much as 90 percent (Long, A. 2006, 32). In the 1960s, earlier RAND researchers recognized the importance of preventing cross-border sanctuary and supply, noting the successful application of the Morice Line to this end. The think tank initiated a series of studies aimed at evaluating proposals for sealing the Vietnamese border. After extensive modeling of cross-border infiltration, one such study by RAND concluded that, "in the absence of a border security system that at least hinders or deters the enemy from determining freely his desired infiltration rates, no model solution leads to conflict termination" (Long, A. 2006, 50). A major anti-infiltration system patterned after the French Morice Line was never fielded in Vietnam and RAND committed no further research to exploring border security after the U.S. experience in Vietnam (Long, A. 2006, 50).

Others have continued to explore physical barriers as a counter to porous borders. Paul Staniland, for example, posited that "sealing off border infiltration routes can turn external sanctuaries into isolated and impotent enclaves" (Staniland 2005-2006, 31). Accordingly, both Israel and India depend on fences and border-defense systems to support internal security. Staniland explained that "Israel claims to have successfully blocked 92 of 93 infiltration attempts from Gaza between January and July of 2005. The Indian government claims that the fence along the line of control has helped reduce infiltration attempts into Kashmir by up to 60 percent" (Staniland 2005-2006, 33). Interestingly, Kuwait is in the process of building a steel fence across its border with Iraq to prevent a spillover of extremism and sectarian violence into its territory (Staniland 2005-2006, 31). These defenses have had their desired effect without sparking unintended and counterproductive effects that would occur with direct military action along or across the same borders.

Physical barriers such as fences, then, have traditionally served as viable counters to porous borders in past scenarios. Similarly, mining was employed widely in the last century in an effort to seal borders or to limit transit across particular areas. Today, however, this approach is tactically inflexible from a military perspective and would most often be politically unacceptable as well. Joint Publication 3-15 explained that air operations could deliver scatterable mines in support of land operations (*Joint Doctrine for Barriers, Obstacles, and Mine Warfare* 2007, II-2, III-1, III-4). Regrettably, aerial mining was the only reference to airpower's applicability to sealing a border or region throughout this joint publication. According to current U.S. policy, U.S. forces are permitted to only use the type of antipersonnel landmines that do not self-destruct for training purposes and for countermine operations in defense of the Republic of Korea (*Joint Doctrine for Barriers, Obstacles, and Mine Warfare* 2007, I-9). Furthermore, mining is widely considered to be outside the acceptable international standards of force implementation. As such, this outmoded method is no longer compatible with warfare in the Twenty-first century. And yet, the problems it addressed in the last century remain. In fact, new BCT employment concepts have increased the likelihood that bypassed forces, or those that have infiltrated porous borders, will pose a real threat to friendly ground units. New ideas are clearly necessary.

During future armed conflicts, advanced technology may give military commanders transformational options with the practical advantages of physical fences and without the political baggage of mining operations and the like. The Twenty-first century strategic environment demands flexible, adaptive, and transformational solutions to the challenges posed by long, uncontained borders. Military commanders would likely look to a spectrum of options with the goal of limiting cross-border flow in certain circumstances and effectively halting infiltration at other times. The same commanders may be unwilling to commit U.S. ground troops in the numbers required when seemingly more pressing missions mandate the use of those limited surface forces elsewhere. Of note, Pakistan's military has reportedly deployed as many as 80,000 troops to counter the flow of

Taliban and al-Qaeda forces across its rugged border with Afghanistan ("Pakistan Willing to Seal Off Border With Afghanistan" 2006, 1). Committing a similar number of U.S. ground forces to the border-security mission would substantially slow the tempo of combat operations elsewhere and runs counter to the goals of vertical maneuver.

Rather than dedicating large U.S. and Coalition surface units to this mission, *AirLandBattle21* proposes the multifaceted, integrated application of air, ground, and space assets. In this way, strategic cordon could be appropriately viewed as another poignant example of the increased interdependence between U.S. Army and Air Force capabilities in the new century. Of course, in some cases, actual physical fences may serve as an important part of the solution. More often, though, commanders will likely call on combat aircraft, unmanned aerial vehicles (UAVs), and space-based surveillance capabilities to erect a less visible fence. In fact, one could quite appropriately draw an analogy between the strategic cordon concept and the increasingly popular "invisible fences" used widely to contain pets within the owner's yard instead of unsightly chain link fences. While the owner can transit the area uninhibited, animals wearing a special collar receive an electrical shock if they attempt to move beyond the borders of the invisible fence. In order to avoid the associated pain, pets learn the lateral confines of the yard and avoid crossing its borders. In the same way, the strategic cordon aims to curb the movement of enemy forces and equipment from one area to another while allowing friendly forces and, when appropriate, non-combatants to freely transit the same regions.

Air and space assets offer substantial potential in this mission area. The most favorable results, however, are likely when these assets can be integrated with a relatively small number of ground forces controlling check points and operating surface equipment such as sophisticated ground sensors. Competent Coalition or indigenous ground forces, which have been sufficiently trained and equipped for the job, may be most appropriately employed in this role. It may, however, be necessary for a small number of American ground forces to serve as a communications interface with airborne surveillance and strike assets. Rather than the traditional mindset of simply "calling in air support", though, the supported/supporting relationship may be more dynamic and more in line with the concept of an air-maneuver force. For example, a UAV operator tracking a potential infiltration might request a closer look from ground forces in the area, asking them to positively identify the suspect vehicles or individuals as enemy forces. Once the friendly ground team identifies these forces as hostile, they might call on a formation of combat aircraft assigned to the area to destroy the target. In other cases, the ground unit would directly attack the infiltrating forces with the aircraft in a supporting role.

As seen elsewhere in this study, much of the technology necessary to perform such a mission currently exists while the transformational concepts and joint doctrine are lacking. Skeptics of a strategic cordon approach featuring air and space assets along with ground troops would likely cite traditional aircraft limitations as impediments and suggest that weather and enemy techniques such as concealment and dispersion would reduce the likelihood of success. While Twenty-first century U.S. air and space assets are far more capable than past platforms, it is true that other capabilities will be necessary to achieve maximum success. This study submits that, as a result of a lack of Service ownership and the aforementioned narrow focus of air interdiction within the Air Force, there has not yet been a firm commitment across the U.S. armed forces to solving the dilemma of porous borders.

As previously established, cross-border activity played a significant role in past wars and was central to the conflicts in Iraq and Afghanistan. More importantly, though, porous borders will continue to impact future medium- to large-scale military operations in substantive ways. The need to cordon adversary forces bypassed by noncontiguous BCTs conducting distributed operations deep inside enemy territory further increases both the likelihood and importance of this new methodology. These challenges would be addressed more holistically in the event that a particular Service took ownership of the mission, developed the requisite comprehensive operational concepts, and seriously committed resources to the development of any additional needed capabilities. Despite DoD efforts to shift to a fully joint, capabilities-based acquisition model, however, the existing process appears at times to remain tied to Service-specific platforms. As a result, serious warfighting issues that cross over Service organic responsibilities may remain underemphasized and under-resourced despite growing evidence of increased, lasting relevance. Consequently, the Services tend toward systems that perform the same traditional functions faster, better, or more efficiently. These improvements are generally Service-specific, conspicuously incremental, and sometimes lacking substantive transformational change. More specifically, because the Air Force does not currently recognize porous borders as its responsibility, the Service is not likely postured to develop capabilities and concepts that would fully enable an initiative such as strategic cordon.

Rather than the existence of platforms driving military capabilities, changes in the operational environment should more appropriately fuel the development of new concepts and required technologies. In practice, it is generally necessary for a service to take a leadership role. For example, the Air Force led early defense efforts in the cyberspace domain, primarily because in many ways its mission is intertwined with the cyber domain and because it has a wide variety of capabilities with which to operate in a dominant role within the cyber regime. As a natural early lead, the Service took steps to formally establish the criticality of the role, helping it to be adequately resourced, and ultimately contributing to U.S. Strategic Command's rightful acceptance of the lead. In a similar way, perhaps the Air Force is a natural early lead for the strategic cordon concept. In the century when the Air Force declared its independence, strategic attack seemed a fitting mission. As the Service embraces interdependence in the new millennium, perhaps alternatives to legacy forms of strategic attack, such as strategic cordon, will take hold.

Traditionally, cordon operations have been viewed as an attempt to *temporarily* seal off a particular area for the purpose of conducting a search. In 2006, the

Air, Land, Sea Application (ALSA) Center released a multi-service tactics, techniques and procedures (MTTP) publication that guides the conduct of search and cordon operations. The document devoted a few short pages to describe Air Force capabilities that might support Army or Marine Corps forces in the conduct of search and cordon operations. An example of this type of cordon is securing a small tactical area such as a series of buildings for the purpose of searching a particular structure for weapons or combatants. Accordingly, the search and cordon multi-Service publication briefly discussed how Air Force assets might use various platforms to provide communication, sensors, and situational awareness in support of the ground force (See *Search and Cordon* 2006). This view of airpower's role in general and the concept of cordon in particular was decidedly tactical and quite limited in nature. *AirLandBattle21*, on the other hand, recognizes the potentially strategic application of cordon operations and places airpower at the center of such an approach. While the Air Force is capable of supporting these small, tactical cordon operations, indigenous ground forces such as rotary-wing aviation assets and hand-held UAVs may be better suited for this limited mission. The application of Air Force resources in this supporting role is akin to the differences between CAS and BAO discussed in Chapter 4. The CAS mission and this search-and-cordon method the ALSA publication addresses both deal with enemy forces already engaged in combat activities. As with BAO, it may be more appropriate and efficient for Air Force assets to cordon large areas to prevent resupply and transit of weapons and combatants *before* these forces can be committed to the close fight. For example, a formation of combat aircraft could assist in efforts to target a few enemy forces engaged in combat with friendly troops or could instead destroy the same forces in large numbers as part of a strategic cordon operation.

Figure 8.1 returns to the previous series of figures in order to provide a visualization of the concept of strategic cordon. In this example, a national border exists to the northwest of the capital. In order to neutralize the cross-border flow of combatants and supplies, a strategic cordon has been established along this border. Similarly, the JFC or JFACC may also choose to establish a cordon within the interior of the enemy territory to inhibit bypassed forces from hampering the distributed operations or logistical supply lines of friendly BCTs. In this way, air assets would greatly enhance friendly surface operations by eliminating enemy freedom of movement across designated zones. In turn, this would enable friendly ground forces to rapidly maneuver with the knowledge that potentially vulnerable flanks were protected by air component cordon operations.

In order to meet with success in the Twenty-first century, strategic-cordon techniques must be both flexible and versatile. They must be adaptable to missions such as eliminating enemy freedom of movement across national borders, restricting combatant lines of communication between towns, and preventing escape into mountainous sanctuaries. Aerial cordon operations may include a mix of lethal and nonlethal capabilities. They must be capable of operating in conjunction with more traditional methods such as physical barriers as well as integrating with small-unit ground teams, sophisticated sensors, and a network

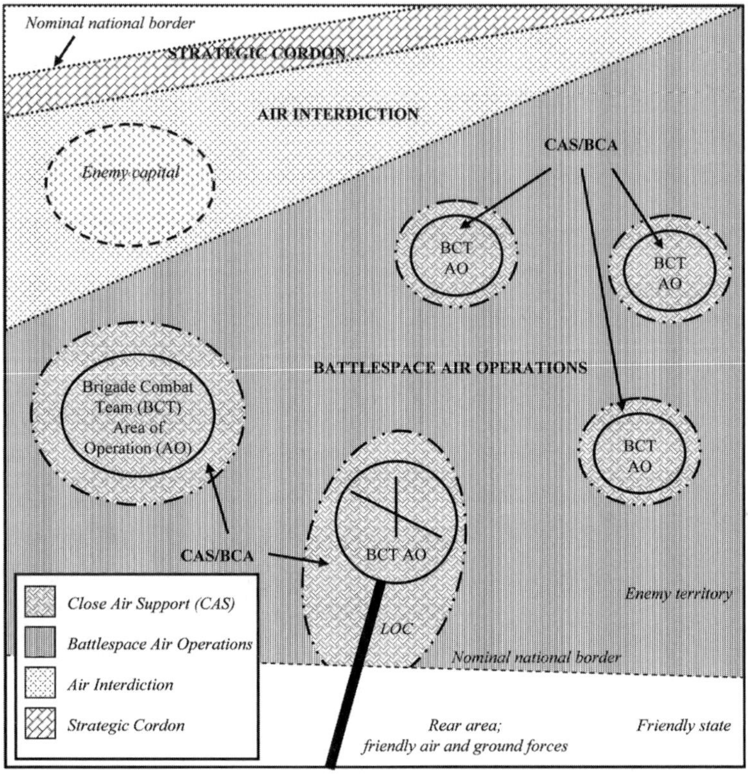

Figure 8.1 Strategic cordon

of border check points. In many cases, aerial forces must allow for safe U.S./Coalition passage across designated lines while simultaneously preventing enemy escape or infiltration. Of great importance, this capability must offer persistence consistent with the level of presence dictated by the mission. If friendly ground-force vulnerability is high, for example, more robust and permanent strategic-cordon capabilities may be necessary.

Merging Improved Technology with New Ideas

Rapid current development in the operational capabilities of modern UAVs, with long loiter times, enhanced sensor suites, and nascent organic attack capacity, holds great promise for the strategic-cordon mission. Once again, though, these improved technologies beg the development of equally innovative concepts commensurate with the changed strategic environment. One such idea would enlist the use of current airdrop techniques and existing platform capabilities to position motion detectors with embedded GPS sensors along specified borders or other required locations. Such a concept would link the sensors to the nearest armed UAV tasked and specifically equipped for border patrol. Flight controllers could monitor several UAVs remotely from duty locations in theater or in the continental United States. When movement is detected along a border, a UAV would automatically cue to the location and cycle through a series of protocols, including Blue Force Tracker technology, to minimize the potential for friendly fire. Via a collaborative environment, the UAV controller would share the real-time feed with a borders joint-fires officer, who confirms the target as hostile and authorizes engagement by an armed UAV or combat aircraft assigned to the mission. If it became necessary to further confirm the potential target was in fact hostile, the special operations ground team tasked to the area would be dispatched to make the determination and then, if required, either eliminate the threat or call on airpower to do so.

Finally, the latest weapons innovations such as directed-energy weapons, lasers, electrolasers, microwaves, particle beams, plasma weapons, and pulsed-energy projectiles may prove applicable to strategic cordon as well. Merging such technology with promising new concepts may enable true transformation in this area. The emerging "area dominance" concept, for example, has driven the Air Force Research Laboratory to explore the viability of employing Dominator, a "cooperative weapon concept designed to loiter for long periods over a target area, fly in constellations, and deploy multiple submunitions against mobile targets" ("Military Worth Analysis of New Concept Weapons", 1). The individual weapons would be disposable assets that either attack a target or self-destruct once their fuel supply has been exhausted. While designed for more traditional purposes, this new technology, once wedded to new ideas such as strategic cordon, could greatly enhance efforts to restrict cross-border activity. To be clear, these ideas are by no means intended as an exhaustive list of options. Rather, they are presented here to illustrate the promise such aerospace technology holds when substantive intellectual energy is committed to challenges such as porous borders and bypassed forces.

These examples may also serve to demonstrate the inherent compatibility between the U.S. Air Force and strategic cordon operations. Formally acknowledging such a role and officially assigning it to the Air Force would arguably constitute a natural extension of existing Service responsibilities. It would be compatible with current aircraft platforms and fielded aerospace technologies while incorporating emerging technologies as well. Strategic cordon would also capitalize on the inherent speed, range and flexibility of air and space assets. Along these lines, the concepts presented in this chapter suggest that the capabilities of air assets would be far more effective and applicable to strategic cordon than to current methods of tactical search and cordon.

A closer look at porous borders indicates this chronic problem has arguably become acute in recent times. When considered in conjunction with the risk posed by enemy forces bypassed by widely dispersed BCTs, the challenges that cordon operations would address are indeed strategic in their potential impact on future military operations. In order to achieve success in this area, however, it is critical that this mission receive adequate attention during planning stages as well as throughout combat operations. Simply stated, border security must be in the plan from "day one." Air and space capabilities, operating interdependently with a relatively small number of ground forces, could enable success in this new mission area. And strategic cordon, as an Air Force responsibility, may prove to be a fundamental merger of technology and ideas that creates new methods of fighting across the spectrum of conflict. It is important to reiterate that, in the *AirLandBattle21* construct, traditional forms of strategic attack remain a potent arrow in the JFC's quiver. Added to his arsenal are the transformational concepts of strategic cordon and counterforce parallel attack. These new options may be appropriate to apply earlier than traditional approaches in some future conflicts and may be more applicable than past methods altogether in other scenarios facing U.S. forces in the Twenty-first century.

Chapter 9
Tactical Airlift Concepts

> Moving this far, this fast, we've eaten up more fuel than anyone ever foresaw, so we have to keep going back to get more. In addition to the fuel, and of course the beans, bullets, and bandages they'll need on the way north, vehicles will break down. Things that are easy to provide when we're back in the garrison, like alternators, batteries and even tires, are like gold out here.
>
> —OIF Army Lieutenant Mike Mullins

Up to this point in the text, this study has offered transformational concepts for integrating airpower with the U.S. Army's new BCT construct. In general, these concepts have been developed under the assumption that the BCT could and would be employed in accordance with the Army's bold and innovative future employment strategy. The purpose of this chapter is to look closely at how these combat brigades can be feasibly employed in a distributed manner and sustained as autonomous fighting units. The 2004 *National Military Strategy* stated that "moving forces into multiple geographic locations will require assured strategic access as well as strategic and tactical lift systems robust enough to conduct and sustain multiple, simultaneous operations" (*National Military Strategy* 2004, 14). With regard to strategic lift, the new BCT construct carries with it a substantial expectation of rapid deployment by intertheater airlift within drastically decreased timelines. As such, intertheater mobility is central to realizing Army transformation. The focus of this study, however, has been the employment of air and ground forces once in theater and not the deployment of personnel and equipment to the theater. This chapter, then, will resume focus on the movement and sustainment of ground forces already established in theater. While other studies have identified serious issues with expeditiously lifting new BCTs to the theater, this chapter will analyze similar intratheater challenges not adequately addressed elsewhere.[1]

The rapid employment of expeditionary ground forces within a particular theater is a critical aspect of U.S. Army transformation plans. The Air Force acknowledged in 2007 that "new operational concepts currently considered by the U.S. Army are increasing the demands for capabilities to deliver personnel and equipment to increasingly numerous and dispersed theater operating locations" (*Concept of Operations for Joint Cargo Aircraft* 2007, 1). A 2002 RAND report on airlift and the brigade combat team more specifically found that "airlift is likely

---

1 See Vick et al. (2002) and *The Army's Future Combat System* (2006) for detailed assessments of intertheater airlift's capacity to deploy BCTs within the proposed 96-hour timeframe.

to play a critical role in high-priority resupply and support operations" (Vick et al. 2002, 117). In order to realize the Army's vision, theater forces and equipment must be moved and sustained with unprecedented speed. Current Air Force airframes, doctrine, and associated concepts were not specifically designed for and do not fully address this new and growing demand.

U.S. joint doctrine divides airlift into two overarching categories: intertheater and intratheater airlift. Intertheater airlift constitutes the strategic movement of forces from one theater to another, such as deploying an Army unit stationed in the continental United States to an overseas location in preparation for hostilities. Intratheater airlift, on the other hand, is conducted within an area of operation. In order to conduct a detailed analysis of airlift associated with BCTs engaged in non-linear, non-contiguous operations, it is necessary to further divide intratheater airlift missions that support ground forces into two subcategories: combat employment and combat sustainment. As previously mentioned, intertheater airlift moves a BCT from its garrison location in the United States, for example, to a theater where combat operations will soon commence. Combat employment missions, on the other hand, enable BCT distributed operations by more specifically moving the brigade from a staging or marshalling area within the theater to a geographically separated area of operation (AO) where the brigade commander plans to actually fight. Combat sustainment, as another subset of intratheater airlift, would in turn resupply and sustain a BCT already established and conducting combat operations in a particular AO. A 2007 U.S. Air Force Air Mobility Command (AMC) document defined combat employment as those missions that "provide operational maneuver for the ground force" by airlifting "combat ready units directly into combat operations for immediate engagement, leading to rapid tactical decisions, shortening durations of battle, and contributing to the rapid disintegration of the enemy forces." According to the AMC document, combat sustainment operations, by contrast, "ensure the timely reinforcement and resupply of combat units" (*Concept of Operations for Joint Cargo Aircraft* 2007, 2).

Intratheater airlift is used to distribute and sustain forces within the operational environment. The Army's visionary concept of vertical maneuver, or distributed operations through the air, describes a new method of employment that is fundamentally dependent on intratheater airlift. In turn, employing ground forces in a non-linear and non-contiguous fashion will have serious intratheater airlift and logistical implications. These changes mandate cross-Service interdependence and forward-thinking concept development. Proponents of vertical maneuver find it to be the quickest and most decisive method of achieving success on the Twenty-first century battlefield. The Army's vision for this new idea, however, demands a rapidly responsive logistical stream and the prompt resupply of armored vehicles, personnel, fuel, water, munitions, and a multitude of other supplies essential to sustain modern combat. Further complicating the issue, the Army concept of rapid maneuver across non-permissive battlespace is generally incompatible with ground convoys, which require secure lines of communication and lengthy travel times. As such, airlift may be the only option capable of meeting this need during critical portions of major combat operations. It is consequently imperative that the Air Force closely analyze the critical role it will play in employing Soldiers in such a manner.

The Army has also become increasingly aware of the new BCT's vulnerability in the area of sustainment. A 2006 Army pamphlet warned that "the operational concept of simultaneous, high-tempo, non-contiguous operations distributed widely throughout the JOA [joint operations area] presents significant challenges to sustainability of the deployed force." The pamphlet further specified that "continuing progress in the revolution in military logistics is critical to achieve the CSS [combat support system] transformation needed to sustain … continuous, large-scale operations" (*The United States Army's Operating Concept for Operational Maneuver: 2015-2024* 2006, 52). One could argue that this logistical transformation will require new technologies, organizations, and concepts. As with the overarching DoD transformation, innovative thought will be critical to integrate disparate technical capabilities into a coherent whole.

Much ink has been spilled of late in various efforts to highlight serious limitations in current plans to deploy, employ, and sustain expeditionary combat brigades.[2] It is not the purpose of this study to refute the findings of these reports. Instead, this chapter explores the art of the possible within the constraints of existing airframes, current procurement plans, and approved acquisition programs. While many analysts have explored the need for new aircraft or additional airframes, such discussion is beyond the scope of this book. Rather than furthering arguments against the logistical feasibility of operational maneuver, this study will analyze current limitations, delineate what can be done within those constraints, and discuss how existing airlift assets can successfully support BCT distributed operations.

The Joint Cargo Aircraft

The Air Force has a very capable fleet of C-130 aircraft designed and tasked to perform the intratheater airlift mission. Some in the Air Force have suggested that, as in the past century, the C-130 can continue to meet the needs of a transformed Army. However, the Army's move to a highly mobile force has raised questions as to the Air Force's capacity to meet the task with a highly overstressed C-130 optimized for the more general intratheater airlift mission rather than the specific employment and sustainment of combat troops. Availability of this airframe has become an issue as well. The Air Force Doctrine Center observed in November 2006 that the "lack of C-130 aircraft has put a strain on an already over utilized airframe. Center wing cracks in the C-130E fleet have reduced the number of C-130 aircraft available for deployment, thus making sortie prioritization within the JOA a hot issue" ("Army Transformation and the Impact on the Mobility Air

2 *The Army's Future Combat System* 2006 and Gordon et al. (2007) are two such efforts.

Forces" 2006, 2). A draft RAND report similarly warned that "the Air Force's intratheater airlift capability, already constrained by the restriction and grounding of more than 80 C-130s … would become 'extremely challenging' if the need for routine sustainment of ground forces increased significantly." The report went on to explain that serious problems with the aging C-130 inventory will soon bring fleet totals well below numbers required to have any reasonable hope of sustaining the new BCT construct (Jordan 2006, 23). These very real limitations, coupled with the fact that the C-130 does not meet many of the specific emerging Army requirements, further emphasize the need for alternatives to the current intratheater workhorse.

A significant step toward meeting the growing Army airlift demand is the joint cargo aircraft (JCA) program. In June 2006, the Army and Air Force jointly signed a memorandum of agreement (MoA) between the two Services that merged two previously separate procurement programs into a single effort to acquire the JCA. This MoA articulated that "the Services will jointly develop a JCA Concept of Operations (CONOPS) that details how the platform will be employed in combat and create or modify existing doctrine to articulate the joint CONOPS for the JCA" ("Way Ahead for the Convergence of the Army Future Cargo Aircraft and Air Force Light Cargo Aircraft Programs" 2006, 4). More specifically, it identified the need for airlift at the tactical level and began to delineate what might be required of such an aircraft: "The JCA is a fixed wing airlift platform that performs airlift missions … predominantly at the tactical level." Significantly, the Memorandum went on to explain that performing this mission would be "most effective when operating the aircraft as close as possible to the end user, and which may require using austere or unimproved landing zones, but also includes more traditional airfield operations" ("Way Ahead …" 2006, 2). As Ann Roosevelt reported in June 2006, the JCA program "grew from considering the kind of battle space the Army faces, in non-linear, non-contiguous areas, where people and supplies need to be moved to meet critical requirements. Airlift keeps troops off the ground, not on the dangerous roads that convoys face" (Roosevelt 2006, 1).

The MoA further directed appropriate agencies within the Army and Air Force to "ensure Service specific requirements, and tactics techniques and procedures (TTP) are established to address JCA CONOPS with particular attention to time-sensitive/mission-critical support" ("Way Ahead …", 4). It is indeed critical that the Army, Air Force, and joint community as a whole fully embrace the vital importance of ensuring the constant, timely flow of critical supplies in a non-permissive environment. Accordingly, the Air Force *Concept of Operations for the Joint Cargo Aircraft* (*JCA CONOPS*), published in August 2007, explained JCA capabilities would "mitigate the anti-access operating environment" by moving "from point of embarkation to point of need, delivering to/from prepared, unprepared, and austere airfields, day or night, without the use of ground-based navigational aids and operating in adverse weather conditions and environments" (*Concept of Operations for the Joint Cargo Aircraft* 2007, 4).

The C-27J Spartan won the contract to become the joint cargo aircraft. As required by Army specifications, the C-27J can land in roughly 1,100 feet of runway and take off in less than 2,000 feet from an unpaved surface at maximum weight. As the Air Force *JCA CONOPS* recognized, these capabilities will "enable bypassing disrupted lines of communication and reducing exposure inherent during ground transportation", replacing convoys in many cases and saving lives in the process (*JCA CONOPS* 2007, 2-3). A maximum speed of 325 knots, service ceiling of 30,000 feet, impressive climb and descent capabilities, redundant systems, and self-protection provisions make for a credibly survivable, rugged airframe capable of performing various airdrop profiles when the enemy threat requires. While the C-27Js maximum cargo load is 25,000 pounds, the practical combat load used in this study, which accounts for multiple considerations during combat operations, is 16,000 pounds. The aircraft can carry two up-armored humvees or two light trucks, can transport dense, heavy loads such as engines and ammunition, and can fit 68 troops when landing is planned or 46 paratroopers during an airdrop mission.

Beyond JCA

The Army and Air Force have clearly found a capable airframe in the C-27J Spartan. The joint cargo aircraft, however, is an answer—not *the* answer—for BCT combat employment and sustainment. First, the sheer volume of lift required by multiple BCTs conducting sustained major combat operations would quickly overwhelm the projected force of joint cargo aircraft. The entire projected Army and Air Force fleet of 145 JCA, for example, could sustain no more than two full brigades in the field simultaneously and, on its own, could meet less than a quarter of a Stryker brigade's lift requirements for a rapid combat employment (see Appendix). Second, and quite significantly, the JCA is not large enough to transport the Stryker fighting vehicle, with or without self protection armor attached, or its comparable future FCS variant as currently envisioned. Third, the C-27J can land on unprepared surfaces but requires some semblance of an airstrip for takeoff and landing that will not be available in all situations. Finally, while conceived as a survivable airframe, there are nonetheless situations where an increased enemy air defense threat might not permit routine JCA airland missions in particular areas. Of course, *AirLandBattle21*'s spectrum of options for relative control in the air is designed to address such a scenario. However, the very real likelihood of a non-permissive environment in future conflicts suggests the need to consider alternatives when organized or sporadic enemy air defense capabilities substantively interfere with JCA operations. None of this is meant to suggest any particular problem with the JCA program. The joint cargo aircraft is indispensable and the C-27J seems a more than adequate fit for the mission. In sum, though, the JCA will prove to be necessary but not sufficient to meet the growing demand of the aggressive new BCT employment strategy.

The Air Force JCA CONOPS held that "the JCA provides operational environment mobility with longer range than Army/Marine utility and cargo helicopters while augmenting current USAF cargo aircraft. It enables tactical airland transport of follow-on forces improving the capability to rapidly reinforce ground units during combat operations" (*JCA CONOPS* 2007, 5). With this statement, the document carved a niche for the JCA between rotary-wing and current fixed-wing aircraft but, in so doing, acknowledged that it would augment rather than replace other existing capabilities. As an example, because the proposed C-27J airframe cannot carry Stryker and other vehicles of comparable size and weight, the platform will be a better fit for sustaining rather than employing a BCT and for smaller loads into austere locations rather than larger loads into prepared fields. These limitations imply the need for capabilities other than the JCA in order to fully meet the requirements of a fully transformed, expeditionary U.S. Army.

Air Force C-130 and C-17 aircraft will accordingly augment the JCA by providing complementary capabilities in these areas. The C-17 can fit as many as three Stryker vehicles but, at the associated 60 tons of cargo, requires a prepared surface when landing in this and most other situations. Interestingly, the Army specified its need for an aircraft that "will be required to move light to medium armor forces, with one or more fully combat capable vehicles or fighting platforms (including crews, fuel, and munitions) loaded internally in a single aircraft" (*The United States Army's Operating Concept for Operational Maneuver: 2015-2024* 2006, 54). The C-17 constitutes the only fielded airframe or acquisition program currently in existence that can fully meet this requirement. Because the C-17 is the workhorse of the intertheater airlift mission, it would most likely be made available in any significant numbers for intratheater airlift only when required to enable the rapid employment of a medium-weight combat brigade. The C-130 is of course dedicated to the intratheater mission and would be available in far greater numbers for direct support to the BCT. Additionally, the C-130 can also transport a Stryker vehicle and has fewer requirements for developed airfields than the C-17. Because the C-130 can barely fit a 20-ton Stryker, however, the self-protection slat armor, which weighs as much as 5,000 pounds, must first be removed. As such, three C-130 sorties are required to transport two Stryker vehicles, and the armor must be reattached upon landing in a potentially hostile environment before engaging enemy forces. The C-130 could of course serve as the primary airlift platform in a medium-threat scenario where Stryker vehicles are required but add-on armor is not. Even in this case, however, some C-17s would likely be required to help offset the BCT's massive lift requirements and to transport other cargo such as fuel trucks that will not fit on any other mobility aircraft. Evolving Army plans for the Future Combat System continue to increase the size and weight of follow-on combat vehicles comparable to the Stryker to the point where they may not fit in a C-130 at all.

It is clear, then, that C-130, JCA, and C-17 aircraft will all serve vital roles in the rapid movement of autonomous brigades. Early disputes between the Services over the need for BCT tactical airlift featured Army proponents arguing for the

JCA with some Air Force skeptics suggesting only the C-130 was needed to perform this intratheater airlift mission. This "either-or" debate missed the point that each aircraft had a unique blend of attributes necessary to adequately support BCT distributed operations. The recent acquisition of new airframes such as J-model C-130s, C-27J Spartans, and additional C-17s suggests that the Services now recognize the need for a combination of interlocking capabilities to meet the need. The same case could be made for the CV-22.

This exciting new airframe is a classic example of a truly transformational technology starving for equally innovative concepts to make it viable. The tilt-rotor technology of Bell Boeing's CV-22 Osprey allows it to cruise at fixed-wing airspeeds while landing and taking off like a helicopter. This unique blend allows the aircraft to fly at up to 275 knots and 26,000 feet while transiting hostile airspace and then land in austere locations with very short unprepared fields or no airstrip of any kind. Stryker vehicles would be too large and heavy for the two-engine tilt-rotor Osprey, but its capacity to carry up to 20,000 pounds of cargo or 24 troops makes it quite useful as a troop and supply transport. Already envisioned for use in the Air Force special operations community, there is clearly an opportunity for the Osprey to allay BCT lift shortfalls as a supplement to the JCA with complementary, unprecedented attributes. Landing with critical supplies in an AO with no airstrip of any kind would be one such example. Additionally, in the event that large numbers of fixed-wing aircraft overwhelm the limited capacity of an airfield supporting BCT operations, the Osprey could land cargo in other less-congested areas. This study is committed to avoiding recommendations for new procurement programs; the Air Force, however, is fielding the Osprey at the present time. Current plans involve use of the emerging fleet of 50 CV-22s exclusively in the special operations role. Consideration should be given to applicability of the Osprey in BCT employment and sustainment when available to perform the mission.

The Army's *Operating Concept for Operational Maneuver*, published in October 2006, expressed the need for capabilities that the JCA and other fixed-wing airframes will not provide:

> Aerial sustainment will be required in greater degree to support the air-ground mobility and agility needed to meet joint force requirements. Advanced SSTOL [super-short take-off and landing] and heavy lift vertical take-off and land (HLVTOL) aircraft will be particularly critical to sustaining operations distributed throughout the entire JOA. Current fixed wing air platforms are insufficient for this role because of their inherent limitations on employment via improved aerial ports of debarkation (*The United States Army's Operating Concept for Operational Maneuver: 2015-2024*, 50-51).

The CV-22 begins to address these expanded requirements. Traditional rotary-wing airframes might serve a role as well. In fact, the Air Force's new combat search and rescue (CSAR) helicopter may offer a blend of vertical takeoff and cargo capacity to serve in some limited cases as a supplement to fixed-wing airlift. While it is crucial for CSAR to remain the primary and central mission of this platform, it would be reasonable for the Service to consider its potential use when available in sufficient quantities to augment the JCA. At the very least, situations where the resupply demands of an isolated BCT become an emergency, the new Air Force CSAR-X airframe could be called on to come to the rescue.

Similarly, the Air Force now has a special operations squadron that flies the U-28, a military version of the Pilatus PC-12. Bruce Rolfsen reported that "the single-engine, turbo-prop plane can carry up to 3,000 pounds of cargo and a handful of passengers." He went on to explain that "the command is interested in obtaining other small planes, such as twin-engine transports, that could deliver gear and people" (Rolfsen 2006, 18). Wray Johnson, a retired Air Force special operator and professor at Marine Corps University, has promoted the acquisition of this weapon system, calling it a "rugged, reliable, survivable and multi-dimensional" aircraft that could perform both the tactical airlift and counter-insurgency mission ("Counter-Insurgency Urgency" 2006, 36). As with the CV-22, it is necessary to emphasize the primary mission of the U-28 would irrefutably remain special operations. However, since the Air Force is already flying aircraft with diverse and potentially applicable airlift capabilities such as the U-28 and CV-22, it would be sensible to explore their applicability to BCT employment and resupply operations.

Airdrop

The joint cargo aircraft, along with the complementary capabilities of these other airframes, will go a long way toward meeting the substantial demands of a transformed Army. However, three potential scenarios remain that require further consideration. First, a sufficient runway may be unavailable while the airlift operation requires fixed-wing assets. If C-17s are required to transport Stryker vehicles, for example, then a prepared surface of considerable length would be required. Second, a prepared surface may be available for BCT airlift operations but, in some situations, it may be inadequate to support an airlift armada. The volume of airlift operations envisioned to support BCT employment, for example, would likely require substantial ramp space, taxiways, and potentially multiple runways. Third, even when operating from an airfield of sufficient size, the enemy air-defense threat may not allow a large number of mobility aircraft landing, remaining on the ground for a certain period of time, and then taking off again.

One seemingly underemphasized and arguably quite applicable method of addressing these and other contingencies would be traditional and transformational airdrop methodologies. In Afghanistan, C-17s demonstrated the applicability of this approach by airdropping 360 tons of supplies to Coalition forces (Tirpak 2007, 29). In a 2005 *Air Force Journal of Logistics* article, Peter Garretson found that airdrop will be increasingly critical given the lessons of recent wars and expected BCT employment techniques of the future:

> Mobility may be less glamorous than dropping bombs, but airdrop has proven vital during Operations Desert Shield; Desert Storm; Iraqi Freedom; and particularly, Enduring Freedom. Airdrop can provide the sole method of resupply in a landlocked theater of operations and is likely to become even more critical in future conflicts given increases in range, speed, and responsiveness. If Enduring Freedom and Iraqi Freedom are decent indicators, we can expect operational plans to be carried out by geographically dispersed, highly mobile land units that consume expendable material as fast as they move. Future wars likely will resemble Iraqi Freedom in that they are waged with comparatively small numbers of U.S. troops moving across vast distances without completely secure supply lines. Future operations will take place in areas where land-based supply lines are highly vulnerable to disruption as experienced during Iraqi Freedom. These attributes make aerial resupply an attractive prospect (Garretson 2005, 39).

Simply stated, airdrop will be essential to success on the battlefields of the Twenty-first century. In fact, operational maneuver will not be possible across the spectrum of likely future scenarios without a robust U.S. Air Force airdrop capability. The Air Force acknowledged in 2006 that "airdrop includes all methods of delivering personnel, equipment, and supplies from an airborne aircraft. This enables commanders to project combat power into areas lacking suitable or secure airfields" (*JCA CONOPS* 2007, 5). Of note, the Stryker can be delivered by both high-altitude and low-altitude methods. In a 2001 *Army Logistician* article discussing tactical airlift and what was then called the interim brigade combat team (IBCT), John McCoy articulated the applicability of airdrop and highlighted an additional benefit of air delivery:

> Prolonged operations require follow-on forces, so resupply by airdrop allows those forces to use a limited-capacity airfield without losing or delaying IBCT supplies at airfield choke points. Substituting airdrop sorties for air-land sorties reduces the exposure of fixed-wing Air Force aircraft to antiaircraft threats near the destination. Also, unless operational security for the mission has been compromised, the exact destination of each airdrop mission cannot be anticipated easily by an adversary, as is possible with air-land missions that must seek out a designated airfield (McCoy 2001, 13).

As the Air Force develops Twenty-first century airlift TTPs to support the BCT construct, it will be necessary for the Service to supplement traditional airdrop methods with more transformational approaches. A very promising technology along these lines is the Joint Precision Aerial Delivery System (JPADS). The Air Force has already seen initial successes with this innovative airdrop system and the concept has been employed during real-world operations (Garretson 2005, 42 and Tirpak 2007, 30). The Service acknowledged in 2007 that JPADS will greatly increase its capacity to meet the BCT's airlift requirements: "Tactical airlift demands are pushing combat delivery further forward to meet the joint requirement to deliver required cargo and personnel where and when they are needed. Expanded airdrop capabilities, like Joint Precision Aerial Delivery System (JPADS), have helped to bridge the capability shortfall by delivering cargo directly to the point of need" (*JCA CONOPS* 2007, 1).

This new JPADS technology provides mobility aircraft with the capability of accurately airdropping supplies from high altitude. JPADS is advancing with new parafoil designs that will allow heavy precision airdrop in the very near future: "JPADS-guided parafoils have demonstrated the capability to drop 10,000-pound packages at 25,000 feet from existing airlift platforms, landing with accuracy within the Army's required 100 meters of circular error probable and an offset up to 25 nautical miles" (Garretson 2005, 41). Under ideal conditions, the C-17 is capable of delivering 40 supply bundles that weigh nearly 2,200 pounds each. Heavier loads up to 40,000 pounds and perhaps higher are becoming a practical reality. High-altitude delivery, standoff capability, and projected costs of up to $30,000 per bundle suggest that the JPADS delivery method will be appropriate for limited, higher-threat scenarios when airland and other lift options prove too dangerous.

Another airdrop method that holds great promise for BCT employment and resupply is the Low Altitude Parachute Extraction System (LAPES). This airdrop system allows an aircraft flying approximately five to ten feet off the ground to deliver up to over 40,000 pounds of cargo. Once released, a drogue chute deploys and pulls the load out of the aircraft. The cargo simply skids to a stop in a relatively short distance. During the Vietnam War, C-130s used the LAPES concept frequently in scenarios reminiscent of the Army's visionary concept in order to fight from remote locations: "The LAPES delivery method was developed years ago to solve the problem of quickly placing loads directly on the ground in areas where it would be unsafe to land the aircraft" ("Low Altitude Parachute Extraction System" 2008). During the siege at Khe Sanh, for example, U.S. Air Force C-130s delivered 567 short tons of supplies on 52 LAPES missions following the decision that the threat from North Vietnamese forces made landing these new aircraft too dangerous (Nalty 1986, 58). In fact, Sean Serfass found in 2004 that "LAPES was employed by the United States military for over 25 years and evolved from a joint Army and Air Force concern for the ability to directly support and supply forces in remote areas under austere conditions" (Serfass 2004, 1). In 1994, the C-17 demonstrated its capacity to employ the LAPES delivery method as well, airdropping single pallet loads as heavy as 41,000 pounds at 130 knots while no more than two feet off the ground (Serfass 2004, 1; "Low Altitude Parachute Extraction System" 2008).

While both the C-130 and C-17 have LAPES capability, the Air Force does not currently employ this useful and cost-effective system. In the aftermath of the debacle in Vietnam, the U.S. Army of the Fulda Gap era did not envision a requirement for such a system while, in the meantime, the Air Force focused less attention on airlift at the tactical level. As the U.S. military transforms to meet the challenges of the Twenty-first century, however, both Services might want to think again. The LAPES delivery method holds great promise for meeting distributed BCT needs. The Air Force, then, would be well advised to seriously consider developing a robust LAPES capability for the C-17, C-130, and joint cargo aircraft. Implementing a substantive LAPES training program for crews in each of these platforms would provide a critical capability as the Air Force works toward meeting the Army's future tactical airlift requirement.

Various methods of airdrop, including JPADS, LAPES, and more traditional approaches, will greatly supplement BCT employment and sustainment efforts when the threat posed by enemy forces requires alternatives to landing aircraft in hostile areas. Counterair concepts introduced in Chapter 6 provide a framework for addressing enemy efforts to interfere with BCT transport operations. In the most high-threat situations, requisite levels of friendly control in the air might at times mandate extreme measures such as fighter escort during airdrop operations. One could envision a scenario not unlike the skies over Western Europe during the Second World War. Rather than B-17s dropping bombs with P-51s overhead, the Twenty-first century version might feature joint cargo aircraft dropping troops and supplies with A-10s as escorts. Some airpower strategists might argue that there are more efficient ways to employ airpower. This might be a valid contention. In some situations, however, it may be one the joint force commander is not interested in hearing. As such, it is important for the Air Force to develop sound concepts and TTPs for conducting such operations in potential future scenarios.

This section has offered a variety of options for sustaining BCTs when the use of ground transportation is not considered viable. It is conceivable, however, that commanders will feel compelled to opt for vehicle convoys even in situations where the threat from enemy combatants on the ground threatens their security. In order to mitigate the threat posed by enemy combatants operating in the gray areas along logistical lines, as has been the case for recent operations in both Iraq and Afghanistan, combat aircraft might again be called on to provide protection to these vulnerable convoys. Such protection may come in the form of armed aerial escort or could more broadly resemble the strategic cordon concept discussed in Chapter 8. The 2004 *National Military Strategy* observed that "force application in multiple overlapping operations will challenge sustainment capabilities. Sustaining such operations requires the ability to support forces operating in and from austere or unimproved forward locations" (*National Military Strategy* 2004, 14). Accordingly, the Air Force and Army will need to develop a "logistical combined arms team" of air and ground assets in this and other ways in order to successfully distribute supplies in support of distributed operations. The interdependence between air and ground forces operating in this integrated manner would constitute the essence of *AirLandBattle21*.

Hybrid "System of Systems" Approach

The previous discussion offered several alternatives to the joint cargo aircraft for employing and sustaining distributed BCTs. When viewed individually, each option offers a piece of the puzzle. In the aggregate, however, these disparate parts begin to fit neatly together into a hybrid "system of systems" approach. As with other aspects of the transformation discussed in this book, the solution is not simply a new system or particular advanced capability. In order to complete the transformation, one must supplement new technology and organizations with new concepts. JPADS and LAPES are two glaring examples of exciting technologies starving for the innovative ideas that will make them transformational. The hybrid "system of systems" approach may be such a concept. Rather than a single system such as the JCA, then, a more holistic answer lies in an approach that integrates different technologies and capabilities into a comprehensive operating concept.

BCT employment and sustainment offers a uniquely complex set of problems. When one airlift method or airframe falls short of meeting a particular need, a second option will pick up where the first left off—each is a link in the resupply chain. Truck convoys, for example, may serve as the primary method of transport in a permissive environment with relatively short logistical lines. The new JCA fleet, supplemented by available U-28/PC-12s, could begin to meet the need when the enemy threatens surface lines of communication and a short airstrip is available. In the absence of a runway of any sort, CV-22s, Air Force CSAR-X helicopters, and organic Army rotary-wing transport could be called on to meet a portion of the airlift need. When an adequate prepared surface is available, C-130s would serve as a vital contributor. In a medium BCT employment scenario, when Stryker vehicles must be moved intact and in large quantities, C-17s would prove to be the optimum airframe. In fact, the shear volume of a combat employment mission suggests each system in the hybrid approach would likely play a significant role. When the threat posed by enemy forces negates the landing option, then traditional airdrop, JPADS, and LAPES could come to the rescue. Finally, the most dangerous scenarios might require dedicated air support such as fighter escort or other suitable alternatives.

In the Twenty-first century, it is critical for forward-thinking ideas to move beyond the Service realm and fully explore the interdependence between the separate Services. The inter-Service hybrid approach offered in this study strives to do just that. Moving from a Service-centric to a joint approach in this way allows for the most efficient and arguably appropriate use of limited assets. Excess Marine Corps MV-22s, for example, might be released, along with Air Force CV-22s not tied up in the special operations mission, to the theater common-user pool to fill Army BCT resupply requirements other assets could not meet. Similarly, excess Army or Air Force JCA platforms in the common-user pool might be used to meet a Marine Corps need for which its organic lift capability is not optimized. As the Marine Corps replaces many of its rotary-wing transport aircraft with the MV-22, Army helicopters could be used in certain situations to meet a niche Marine requirement. Similarly, Marine Corps KC-130s may prove very useful in the unique mission of transporting fuel to meet the needs of non-contiguous, geographically-separated Army BCTs. An Air Force C-130 may in turn backfill the Marine Corps lift requirement previously filled by the KC-130. Moving beyond Service-centric solutions to joint interdependence would round out the hybrid "system of systems" approach and further increase the likelihood that BCT lift demands can be met in a timely and uninterrupted fashion.

What remains is a tangible analysis of this tactical airlift system's capacity to provide for simultaneous BCT employment and sustainment. While a literally massive undertaking under surge conditions, calculations discussed in detail in the Appendix indicate that the existing and projected fleet of available fixed-wing Air Force and Army aircraft can employ a single Stryker brigade to a combat AO in a 24-hour period. In accordance with the assumptions discussed in the Appendix, the combat employment of a Stryker brigade weighing 15,000 tons would require an estimated 253 aircraft flying a total of 759 missions. While intended to allow for simultaneous flow of intertheater airlift, BCT intratheater movement analyzed in this study would likely require up to 20 percent of the C-17 fleet, 87 percent of Air Force C-130 aircraft, and 100 percent of the project JCA airframes. As such, the simultaneous employment of two or more brigades by air does not seem feasible. However, current and projected Army and Air Force assets could employ a second medium-weight brigade over a 48-hour period immediately following completed movement of the first brigade using many of the same aircraft. During and after the employment of these two brigades, a division-sized force of four geographically separated, autonomous BCTs conducting medium- to large-scale major combat operations could be sustained indefinitely by air under fairly routine conditions. A total of 41 aircraft would fly 81 daily missions to sustain one brigade by air while 164 aircraft would fly 324 sorties to provide sustainment for all four brigades (see Appendix). While it would be feasible for these airlift assets to sustain a larger number of Stryker brigades in the field simultaneously, such a scenario would not reasonably allow for the rapid aerial movement of these distributed brigades or the employment of an additional brigade while sustaining those already in the field.[3]

The massive airlift effort that Army operational maneuver will require draws attention to other demands on the same assets. While this study has purposefully focused on the intratheater issue, the necessary use of multiple C-17s within the theater further complicates the monumental task of rapid BCT deployment. Additionally, the requisite intertheater and intratheater movement of Air Force assets would not allow the complete dedication of the airlift fleet to the deployment, employment, and sustainment of BCTs. Viewing the issue from an *AirLandBattle21* perspective, it would be inappropriate to characterize Air Force lift requirements as "competing" with Army needs. Instead, these demands are

3 It is important to note that these calculations do not include organic Army rotary-wing assets. The availability of Army helicopters in substantial numbers during an actual airlift operation would provide some relief to other over-stretched mobility assets and potentially mitigate to a certain degree the serious challenges identified in this study.

complementary, as both play a critical interdependent role in the success of the war on the ground. The Army's distributed operations concept simply would not be feasible without the presence of combat airpower in adequate numbers. The positioning of these forces requires substantial airlift. A war plan calling for the simultaneous deployment and employment of large numbers of BCTs, then, would greatly reduce the availability of the airpower required for the survivability and ultimate success of the operation.

These logistical realities will force joint force commanders during future operations to make exceedingly difficult decisions along these lines. Such constraints will require air component commanders and their planners to make hard choices as well. The airlift constraints revealed by this study suggest that, when the Army employs operational maneuver, it may prove necessary to commit much of the air component's combat airpower and airlift to that end. Simply stated, missions beyond the realm of interdependent operations may not be feasible at the initiation of many future combat actions. More specifically, it is not likely that the current and projected airlift fleet can logistically support a theater strategy involving both traditional forms of strategic attack and the combat employment of multiple geographically separated BCTs at the outset of combat operations. Missions that complement the distributed operations model, such as battlespace air operations, close air support, counterair, and perhaps alternatives to strategic attack such as counterforce parallel attack, might instead be the necessary central focus of the air effort.

Analysis further indicates that employing and sustaining autonomous BCTs in hostile environments would generally require local air superiority at a minimum. While multiple considerations dictate airlift methods, one might select airdrop as a primary method of BCT movement and resupply when this form of relative control in the air has been achieved. In extreme cases, fighter escort may be one method of gaining this local air superiority. Because of inherent vulnerabilities associated with landing and taking off in a hot zone, some commanders might opt for general air superiority or even local air supremacy prior to selecting airland operations as the primary delivery method, particularly when multiple aircraft would likely be on the ground at the same time. The point here is to highlight once again that the enemy air defense threat is not a static condition and that commanders can make choices that will mitigate the effectiveness of that threat and the impact it would have on airlift operations.

Offering a brief scenario may provide some clarification to the multifaceted aspects of the hybrid "system of systems" approach. In this example, a Stryker brigade plans to employ operational maneuver to establish a remote AO with an established airfield. The air component quickly achieves necessary levels of control in the air and advance units secure the airfield, allowing airlift to rapidly move the BCT into place. The presence of a prepared surface and airfield environment allows multiple C-17 missions to land and rapidly offload the brigade's Stryker vehicles. JCA and C-130 missions lift the bulk of the brigade's personnel and other cargo. This BCT is fully in place and prepared to initiate combat within 24 hours.

A second brigade begins to establish its AO the following day in an equally critical but more austere location. Rather than landing and offloading Stryker vehicles, C-17s instead apply newly approved LAPES procedures to move the second brigade's fighting vehicles into place. The existence of a short, unprepared field allows C-27Js to rapidly position personnel and other critical supplies. Meanwhile, multiple C-130s employ JPADS to accurately airdrop large quantities of equipment. Additional C-130s airdrop Stryker vehicles into the AO while still others drop the slat armor that Soldiers affix to the vehicles once on the ground. Available U-28/PC-12s and a few CV-22 Ospreys supplement the airlift effort, but CSAR-X helicopters did not participate as they were committed elsewhere.

The brigade commander in this second AO finds that one of his battalions can be positioned with some level of immediacy, but that it takes some time for the remainder of the brigade to arrive. This, however, might not be such a bad thing. In fact, such a scenario brings the entire *AirLandBattle21* approach full circle. Anticipating this eventuality, joint planners might have purposefully selected this particular BCT AO to bait the enemy into an engagement once one of the Stryker battalions is in place and fully combat ready. As airlift moves the remainder of the brigade into place, the enemy force may perceive vulnerability and choose to attack. As the enemy maneuvers for attack, they would fall into the preplanned trap by unwittingly moving into the CFACC's desired engagement zone. In this scenario, the newly positioned battalion may serve as the supporting force while an air-maneuver force violently attacks the enemy unit. Meanwhile, the BCT continues to expeditiously build combat power to full brigade strength and, when appropriate, attacks the weakened enemy force with airpower in support.

Conclusion

A central theme in this text has been to encourage new ideas for conducting warfare in the Twenty-first century. The notion of distributed operations through vertical maneuver meets that criteria. New concepts embodied in *AirLandBattle21* may very well enable these new ideas in the areas of counterair, counterland, and alternatives to strategic attack. To be fair, though, air and ground forces could "muddle through" to a certain extent without the benefit of the newest concepts for winning the war in the air and on the ground. Although more painful, more dangerous, and far less efficient, the objectives of the operation would, in many cases, be achieved nonetheless through more traditional means. If serious issues associated with sustaining the BCT from the air go unresolved, however, failure under less than the most ideal conditions will become far more likely, regardless of the application of innovative concepts elsewhere.

The death of Tactical Air Command at the hands of Air Force Chief of Staff General Tony McPeak coincided with an erosion of missions such as tactical reconnaissance and tactical airlift. The continued Twentieth century prioritization of strategic attack over the counterland mission was arguably matched by a concurrent emphasis on strategic airlift at the expense of dirtier missions such as sustaining Soldiers already engaged in the fighting. As the term *tactical* lost favor and was largely removed from the Air Force lexicon, many associated missions seemingly suffered a similar fate. When the Air Force looked beyond the tactical reconnaissance mission in favor of strategic assets such as satellites and U-2 aircraft, the Army predictably moved in to fill the void, growing an armada of unmanned aerial vehicles to perform what they viewed as a vital yet neglected mission. Today, the largest air force in the world belongs, interestingly, to the U.S. Army. If the Air Force does not move aggressively to meet the Army's requirement for tactical airlift, the "new Army Air Corps" may address the issue in a similar manner. The rise of the Air Force Predator UAV has breathed new life into the important Air Force mission of tactical reconnaissance. It also suggests the need for similar emphasis on tactical airlift capabilities. The advent of innovative *AirLandBattle21* concepts to accompany new technology such as the JCA and JPADS will help to enable BCT distributed operations and, in so doing, complete inter-Service transformation.

Chapter 10
Engaging the Debate and Completing the Transformation

> Air Force doctrine development is never totally complete—it is a continuous work in progress. We must remain aware of the lessons of the past—alert and receptive to future technologies and paradigms that may alter the art of air and space warfare.
>
> —*Air Force Basic Doctrine*

In the Twentieth century, the most zealous airpower advocates may have on occasion been guilty of overselling the capabilities of airpower. At times, they encouraged a perception analogous to the story of the boy who cried wolf and, in so doing, invited a level skepticism as a result. In the Twenty-first century, the wolves may truly hold the proverbial village at risk, and the transformational capabilities resident in both the Army and the Air Force are required to save it. It is paramount now more than ever, then, that Airmen clearly articulate their capabilities and their limitations to their Army brethren as they relate to future joint military operations. As such, a study advocating seamless air and ground integration would be incomplete without addressing limitations to airpower and associated arguments that might be raised in opposition to an approach such as *AirLandBattle21*. Even the highly-sophisticated and technologically-advanced U.S. Air Force faces challenges as it integrates with the BCTs of the Twenty-first century.

As air and ground forces become increasingly interdependent, it may be necessary to begin by emphasizing the fact that airpower applied appropriately can compensate for risk on the ground but cannot always save a bad plan. The aforementioned use of organic rotary-wing Army aviation in lieu of more survivable and capable fixed-wing air component assets against Saddam's Republican Guard in 2003 demonstrated the criticality of ensuring that such planning fully involve operationally savvy air and ground planners. Analysis of the planning for Operation Anaconda serves as further reminder that, in the Twenty-first century, ground operations planned independently of air component representation are fraught with danger. The fact that airpower came to the rescue in the Shah-i-Kot Valley may have reinforced a lingering perception among some Soldiers that the CAS "911 force" can salvage even the most dire, self-induced emergencies. Such a belief could court disaster in the resource-constrained, interdependent environment confronting joint forces in the new century. An Operation Anaconda planned jointly from the beginning, on the other hand, may have integrated air and ground operations and featured the application of airpower in ways similar to the concepts furthered in this study. Fully including imbedded air planners in a similar manner during OIF planning may have led to far greater success and fewer losses during March 2003 attacks on the Iraqi Medina Division as well.

As the Army transforms toward the BCT construct, the Air Force has begun to substantially increase Service representation at the brigade-level. This trend is both encouraging and absolutely critical. With the advent of the distributed BCT operations model, it is crucial that Air Force officers participate early and often as equal partners in the planning activities at the brigade level and above. Traditionally, air liaison officers have focused primarily on the tactical application of airpower in support of ground forces and were generally expected to simply plan and execute the CAS mission. In the Twenty-first century, Air Force officers imbedded at the brigade level and above must be well-versed in airpower theory, doctrine, and capabilities and possess the knowledge to espouse airpower's capacity to serve in both supporting and supported roles. They must not only coordinate tactics and close air support but must be equally skilled at planning at the operational level and executing new missions such as battlespace air operations. Additionally, it is interesting to note that legacy terms such as *air support operations group*, *air support operations squadron*, and *tactical air control party* have arguably institutionalized to a certain extent the idea that air exclusively serves in a supporting role and that the imbedded Airmen's focus should remain exclusively at the tactical level. In an effort to complete the transformation, consideration should be given to a new vernacular reflecting new organizational constructs that complement the concepts advocated in the study.

In the Twentieth century, fratricide proved to be a seemingly unavoidable eventuality when aircraft routinely expended weapons in the vicinity of friendly ground forces. Potential critics of *AirLandBattle21* might highlight this traditional limitation, suggesting that efforts to further the integration of air and ground forces would inevitably increase the likelihood of fratricide as well. Army Colonel Pete Curry, for example, voiced concern about "blue-on-blue" incidents, explaining that the counterland concepts advocated in this study would be most feasible "only when the ground commander is very comfortable with the idea that there's not going to be any 'friendlies' in the location, everyone there is a bad guy, and when the targets are easily identifiable" (Curry 2006). It is true that application of this concept would require advanced parties of the BCT such as forward patrols and reconnaissance elements to clearly communicate their position and fully understand the confines of their pre-approved operating areas. However, restrictive measures such as no-strike targets and no-fire areas, already in use in various settings, would be compatible with non-linear distributed operations and would diminish the possibility of blue-on-blue incidents (*Kill Box TTP* 2005, 13). The essence of *AirLandBattle21* is a tightly integrated effort between the air and land components during both planning and execution. Fully integrated Air Force and Army planners working seamlessly from concept-development to concept-implementation as advocated here would decrease rather than increase the likelihood of fratricidal attacks from the air. New command-and-control concepts, such as the joint operations center (JOC) or joint warfighting headquarters (JWFHQ), are needed as well and would greatly facilitate efforts toward joint planning and integrated operations.

It is true that the counterland concepts advocated in this study require commanders to assume a certain level of risk in this area. When applied in accordance with the guidelines discussed here, though, the risk with these new counterland concepts would arguably be no greater than historical approaches to attacking enemy forces. Fratricide is in fact more likely when the air component is called on to "save the day" in an ad hoc fashion during a chaotic emergency on the battlefield then in a fully-integrated, deliberately-planned, and well-orchestrated air-ground operation. Finally, the *AirLandBattle21* construct does not supplant the use of CAS and positive control measures when friendly troops are in close proximity or in contact with enemy ground forces. The traditional concept of close air support remains fully intact under this new approach. The substantive change to CAS employment would be transforming the fire support coordination line into a new fire control measure more compatible with the non-linear, non-contiguous nature of future BCT employment. A functional JOC or JWFHQ construct would facilitate the employment of a more dynamic approach.

Furthermore, as is the case with full transformation, advanced information technology has drastically improved the interface between Twenty-first century air and ground forces. Improved identification-of-friend-or-foe (IFF) technology such as Blue-Force Tracker are interoperable between the Services and, as such, will assist in decreasing the risk of friendly fire, particularly once such capabilities are fully fielded across the fleet of Air Force combat aircraft and Army ground equipment. The advent of network-centric capabilities and the mass proliferation of mutual communication technologies such as data link will provide a quantum leap in airpower's capacity to accurately identify the location of friendly forces, even in tomorrow's fluid battlespace. This ability, combined with greatly improved targeting information and the drastically increased accuracy of precision weapons, will further mitigate the risk of fratricide from the air.

Beyond concerns about friendly fire, other *AirLandBattle21* skeptics might resurrect the age-old adage of a "fair weather Air Force" and argue that airpower's effectiveness is still greatly diminished when weather conditions deteriorate. This critique might suggest that the aggressive concepts furthered in this study would further complicate this problem. Naysayers might postulate that, just when the BCT is most vulnerable and most dependent on airpower, the bad weather will arrive and the air component will fail to deliver. While weather certainly dampened the anticipated effects of airpower at times in the last century, transformational technologies and vast improvement in all-weather airpower capabilities make such incidents far less likely in the new millennium. Furthermore, it bears mentioning that interdependence between Twenty-first century air and ground forces is a foregone conclusion. *AirLandBattle21* has not created a dependent situation. Instead, the transformational concepts herein were created to address this preexisting condition and, regardless of weather, aim to fill the critical intellectual void in this area.

Curry reflected deep-seated concerns about weather when he explained that "the ground commander is only going to give you a sector of his zone or section when … he doesn't depend on the weather. It's just nice to have that you're on the right flank, for example—that you can be given a zone—but he can't depend on it because, when the weather goes bad, now he has an exposed flank" (Curry 2006). It is true that, in certain CAS scenarios when friendly troops are in contact with enemy ground forces, inclement weather will remain a challenge, albeit less prohibitive than in the past. Interestingly, though, it is the CAS mission itself and its associated positive control requirements, such as positive target identification and affirmative clearance to expend, that become so much more difficult in bad weather. Notably, *AirLandBattle21* substantially reduces cases where CAS is unnecessarily applied and replaces it with the more appropriate, less restrictive mission of BAO. With the advent of a wide array of all-weather precision weapons that can precisely attack any assigned target, such as the Joint Direct Attack Munition (JDAM), and continued improvements in targeting intelligence, which yield evermore accurate coordinates, weather will have a much reduced impact in a BAO environment. As a result, Twenty-first century technology combined with the transformational concepts in this study should mitigate—rather than increase—concerns about weather during future military operations.

The next indictment that critics may level against *AirLandBattle21* might find fault in the spectrum of relative control in the air. Proponents of air superiority and air supremacy might find local air superiority dangerous and even reckless. It is true that an approach calling for only the minimum required level of control, such as local air superiority, would require a level of risk that runs counter to traditional airpower thought. If the joint force commander or JFACC choose to reduce such risk, *AirLandBattle21* provides a menu of options for higher levels of control in the air. Just as the new BCT is more adaptive and modular than its Twentieth century predecessor, the new relative air control model is similarly flexible. In fact, this debate between old and new captures the best of the Defense Department's increasing efforts towards interdependence in the joint fight. Today's brigade combat team is leaner, lighter and more mobile than yesterday's divisions; airpower employed in accordance with these transformational counterair concepts invites the same opportunities. The transformed Army prioritizes audacity and mobility above caution and mass. In the same spirit, achieving temporal counterair effects may introduce some risk but will likely enable far greater results in the war on the ground.

Still other skeptics, intrigued by the allure of the Army's Future Combat System, might contend that sophisticated FCS technology and all-encompassing capabilities will diminish the Army's dependence on its sister Service. First, even the most optimistic estimates put initial fielding of full FCS at 8-10 years and complete integration at over fifteen years. Clearly, then, with the Army transition to brigade-level modular forces well on its way, interdependence with the Air Force is a reality in the short- and medium-term at the very least. Second, there remain some fairly substantial fiscal and technological hurdles to overcome on the road to FCS implementation as currently envisioned. Colin Gray found in 2005 that "the U.S. Army's transformation plans are awash with prudent words on the many dimensions of future conflict, but at its core lies a drive to acquire an all but unaffordable (at $92 billion plus) Future Combat System, consisting of a network of fifty-three vital technologies, nearly all of which are technically unproven at this writing" (Gray 2005, 29). Finally, even the full realization of the promise of the Future Combat System will still require the Air Force to perform critical roles in counterland, counterair, and tactical airlift and arguably in ways quite similar to those discussed in this study.

Commentators may instead look at *AirLandBattle21* in the aggregate and suggest that it offers nothing really new, concluding that the concepts presented are not truly transformational. It is true that some ideas furthered herein, such as the air-maneuver force, have begun to emerge elsewhere in one form or another (See Givens 2002). However, this book has introduced new concepts, to include a spectrum of relative control in the air, local air supremacy with general air superiority, battlespace air operations, battlespace coordination area, counterforce parallel attack, strategic cordon, and a hybrid "system of systems" approach to the BCT tactical airlift requirement. Figure 10.1, which summarizes the major findings of this study, draws attention to the novel nature of *AirLandBattle21*.

Some ideas developed in this study were nonexistent before being presented here. Others took their roots from preexisting initiatives across the Air Force community as well as those recently emerging in Army circles. In general, however,

Spectrum of relative control in the air: A spectrum of control in the air, featuring general air superiority with local air supremacy, provides a framework for efficient use of limited air assets, allowing air strategists maximum flexibility to plan for only necessary levels of air control in different areas across the theater.

Battlespace air operations: A new counterland mission that addresses "gray areas" between BCTs and the current doctrinal void between CAS and AI.

Battlespace coordination area: A fire support coordination measure more compatible with non-linear BCT distributed operations than the traditional FSCL.

Air as a maneuver force: Concept that allows airpower to initially serve in a supported role with friendly ground force action compelling a favorable enemy response.

Counterforce parallel attack: An alternative to strategic attack that attacks military targets at the tactical, operational, and strategic levels of war.

Strategic cordon: An alternative to strategic attack that provides a range of options to the JFC to seal off or cordon large areas on the surface such as national borders and geographic regions with airpower as a central feature.

Hybrid "system of systems" approach: Joins JCA, C-130, C-17, CV-22, USAF CSAR-X helicopter, U-28/PC-12, traditional airdrop, JPADS, LAPES, and fighter escort options into a single concept designed to address serious BCT demands on tactical airlift.

Figure 10.1 The basic precepts of *AirLandBattle21*

these efforts have not synthesized such thought into a coherent and comprehensive airpower theory. If anything, *AirLandBattle21* modifies the traditional Air Force viewpoint by conceptualizing the battlespace from the BCT's perspective on the ground rather than from 25,000 feet above. The results stand in stark contrast to the Service-oriented airpower concepts from the last century.

Similarly, much has been written about modern airpower, and substantial literature is available on the brigade combat team as well. Very little exists, though, that seriously analyzes the separate entities of airpower and the BCT fighting as a seamless whole. No effort since perhaps AirLand Battle of old has pulled together concepts into a single, holistic strategy for integrated air and ground operations. Rather than following Twentieth century trends escalating toward air dominance and away from the land battle, this study uniquely advocates an interdependent counterair approach directly tied to realities on the ground. Theories for strategic attack were widespread in the last century but none were inextricably linked, as in *AirLandBattle21*, to a comprehensive approach for integrated air and ground operations. There have been detailed studies on the intertheater challenge posed by expeditionary BCTs but, once again, very little that aims to fully address the intratheater issue as this book has done.

Naysayers might instead find the concepts espoused within these pages too prescriptive in nature. On the contrary, the intent has been for these ideas to simply inform future strategists and assist them in their efforts to develop plans for future wars involving air and ground forces. By no means intended as a joint planning checklist, *AirLandBattle21* could instead be viewed as an alternative framework replete with ideas germane to Twenty-first century military operations. Of course, the actual situation confronting future planners may be different from the assumptions that formed the baseline for these concepts. Guidance from the joint force commander and component commanders will drive the planning effort; these concepts may assist their staffs in meeting the commander's intent. The goal would be for *AirLandBattle21* concepts to inspire innovative future thought, to serve as a worthwhile reference, but ultimately to return to the shelf as the actual plans for actual contingencies take form.

Still other skeptics might find *AirLandBattle21* too reminiscent of its Cold War predecessor. To be fair, some counterland concepts in this study are quite similar to Starry's concept of attacking Soviet follow-on forces in the Fulda Gap with airpower. Both approaches venture beyond the narrow scope of close air support toward methods that attack the enemy's fielded forces in a more practical and efficient manner. AirLand Battle of the last century comes with baggage, however, as it was clearly oriented toward the Army and confined airpower to its traditional supporting role. Its opponents contend to this day that the fundamentally ground-centric precepts of AirLand Battle subjugated the Air Force to a status akin to "maidservant of the Army", relegating airpower to little more than "airborne artillery." *AirLandBattle21*, on the other hand, aims to transform previous Service-centric mindsets into a fully integrated approach. It may be worth noting that, during the same timeframe as AirLand Battle of old, the Air Force continued to develop its own Service-oriented methods of employing airpower independent of ground forces. The concepts of *AirLandBattle21* move beyond these legacy mindsets, aiming to place these previously divergent approaches on converging paths. These new ideas will allow air and ground forces to operate interdependently, in the context of a dynamic supported-supporting relationship, and will supplement the Twentieth century countervalue approach to strategic attack with counterforce parallel attack.

Beyond this comparison to AirLand Battle, others may argue that *AirLandBattle21* is not necessary, as all pertinent concepts for air and ground operations are already captured in joint doctrine. These critics might contend that, while AirLand Battle of the 1980s was appropriate because joint doctrine did not exist then as a mature discipline, the Twenty-first century sequel would not be necessary in the era of voluminous joint publications. It is true that joint doctrine serves a vital role and a noble goal. Joint doctrine, though, is written by consensus and is often the product of the kind of consensus that follows heated and sometime debilitating debate. Inter-Service wrangling over joint terminology and new concepts further slows an already laborious review and revision process. Doctrinal battles over the concept of *effects-based operations* would provide just one illuminating example. The current method, then, leads to text that lags the type of substantial change confronting the joint community at the outset of the Twenty-first century. Frequently cut during such a system are transformational ideas not unlike those developed in this study. As such, the potentially contentious methodologies espoused herein would not be fully compatible with the pace or divisive nature of the current joint doctrine process. While arguably appropriate for Service and joint doctrine, none of the major concepts discussed in this book have yet found their way into joint doctrine in any recognizable form.

Rather than challenging the ideas developed in this study, the debate might instead focus on what is missing from *AirLandBattle21*. A new command-and-control architecture, for example, was mentioned as necessary in the text but was not addressed extensively in this book. Similarly, intertheater airlift is an issue critically important to the new BCT construct but not addressed in this study. As discussed in the previous chapter, other reports have thoroughly analyzed this aspect of Army transformation. As such, it was more necessary and fitting to similarly scrutinize the many intratheater challenges that have not received adequate attention. The ideas, assumptions, and calculations incorporated into that endeavor were designed in full recognition of the substantial intertheater demands simultaneously drawing on some of the same mobility assets. The ultimate objective was to determine what tactical airlift could be accomplished within the constraints of concurrent strategic mobility requirements. This approach of isolating the intratheater problem seems to suggest that modest BCT employment and sustainment can be accomplished while the intertheater flow of forces continues in a moderate fashion. A study that builds on this conclusion by more closely analyzing the interplay between these two critical aspects of Army transformation would be warranted.

Intelligence, surveillance, and reconnaissance (ISR) is another important area not specifically addressed in this study. ISR in one form or another has been a vital aspect of warfare throughout history and has become even more critical in the new century. In fact, certain concepts advocated in this study, such as the "hammer and anvil" approach and enticement strategy discussed in Chapter 5, place high priority on actionable information about enemy disposition and intentions. Lighter, smaller, and geographically-separated BCTs conducting distributed operations will in many cases be far more dependent on ISR than were the large contiguous forces of the past. The Future Combat System will feature a multitude of various intelligence, sensor, and UAV capabilities. The proliferation of Air Force UAV platforms will increase joint ISR capacity as well. New concepts and organizational constructs that integrate emerging Service capabilities and extraordinary advances in information and ISR technology into a single, seamless ISR framework would be a welcomed sequel to this study.

A final critique against this study might contend that, while potentially useful, the ideas herein amount to little more than tactics, techniques, and procedures (TTPs) for air and ground forces conducting combat at the tactical level of war. Forward-thinking tacticians might find some utility to these approaches on the battlefields of the future. However, the objective of *AirLandBattle21* has been to offer operational concepts which, if implemented as intended, would yield lasting effect across the levels of war. The bulk would be most frequently applied at the operational level and, in the aggregate, would have substantial strategic impact. These concepts might find their way into a range of doctrine, from TTP manuals to overarching doctrinal documents, with tactical, operational, and strategic implications to the planning and execution of future military operations.

The military tends to embrace technological change and adjust organizations accordingly, but it is often slow to institutionalize new ideas. Barriers to change such as bureaucratic inertia and military organizational culture often interfere with the acceptance of innovative methodologies. There is no premium on transformational thought; good ideas are certainly out there. This book has synthesized and perhaps added fresh approaches to a pre-existing body of literature. Concepts presented in the past have at times been intellectually stimulating but not necessarily developed to realistically enable full implementation in an environment that is slow to accept intellectual change. The intent with this study was to go a step beyond theory by presenting ideas in a practical manner, facilitating a seamless transition from concept development to concept execution and allowing for their actual application in future military operations. To fully integrate the interdependent air and ground forces of the Twenty-first century, transformational concepts such as those presented in this study must be accepted and implemented in the Army, Air Force, and across the joint community.

In the foreword to the Air Force doctrine document on operations and organizations, the Air Force Chief of Staff, General Michael Moseley, observed the following:

> We are in the middle of a long, shadowy war that will be punctuated by episodes of full-up, conventional operations. Many different types of operations will occur simultaneously; tempo and objectives may shift rapidly. We will be challenged to adapt well-practiced tactics and to think in new ways to solve unforeseen challenges. As the war evolves, so must we. Some aspects of warfighting, however, will remain constant. Our doctrine captures these enduring aspects. Our success in meeting the challenges of this rapidly changing world depends on our understanding and applying our doctrine (*Operations and Organizations* 2006, Foreword).

Many of the precepts furthered in this study, enabled by new perspectives, arguably rise to the level of enduring doctrinal concepts. The challenge will be to incorporate these and other forward-thinking ideas in a way that transforms doctrine and reflects the changed strategic landscape of the Twenty-first century. In a 2006 paper for Joint Forces Command, Retired Army General Gary Luck acknowledged that "we can't simply transfer the old doctrine and TTP that was so successful … when we had a much larger force structure and different environment … Old doctrine (overwhelming force) and TTP simply won't work in today's complex environment and force structure. We need new doctrine and new TTP" (Luck 2006, 7). The ideas in this book would constitute a reasonable place to start.

All told, *AirLandBattle21* is more a journey than a destination. While offering a cohesive set of concepts in the areas of counterland, counterair, alternatives to strategic attack, and tactical airlift, the list is by no means exhaustive. As mentioned, much is left to explore in the areas of command and control, ISR and intertheater airlift. More fundamentally, the Air Force must coalesce transformational ideas like those in this study into a body of doctrine that forms a new baseline for airpower theory in the Twenty-first century. It must seriously reconsider Twentieth century conventions that relegate the increasingly critical counterland mission to a distant third behind counterair and strategic attack. The purpose of this text has been to propel airpower strategists in that direction by engendering the discussion, debate, and intellectual energy that will generate a comprehensive body of ideas integrating Twenty-first century airpower with the BCT. As such, it would be appropriate for some elements of *AirLandBattle21* to inspire the genesis of new Service and joint doctrine. In an official 2006 Air Force document, the Air Force Chief of Staff submitted that "doctrine … must be updated to reflect lessons learned from recent and ongoing operations while taking into consideration Army modularity and technological advances" ("Air Support Operations Center Enabling Concept" 2006, 5). This single sentence cites the need for new ideas to accompany new organizations and new technology. Air Force Major Thomas Svetecz similarly found in 2001 that "technology itself is not decisive; it is the effective integration of new technology into strategy and doctrine that is revolutionary" (Svetecz 2001, 12-13). Developing concepts that complete the technological and organizational transformation has been the motivation for *AirLandBattle21*.

Appendix
Airlift Calculations and Assumptions

Airlift calculations are based on an assumed sortie length is 1.5 hours each way with a total of three hours on the ground in both locations for on-load, off-load, and required maintenance actions. This 6-hour cycle would allow four sorties per day per aircraft under ideal conditions. In order to allow for combat losses, battle damage, aircraft with significant systems malfunctions, and other maintenance delays, the calculations use an average sortie rate of 3.0 missions per day to approximate surge conditions during BCT combat employment. For BCT combat sustainment, JCA and C-130 aircraft would fly an average of 2.0 sorties per day and, with fewer numbers to offset maintenance problems, C-17s would fly an average of 1.0 sortie per day (see Figure A.1).

A Stryker brigade is approximated at 15,000 tons for combat employment. This was assumed to include 320 Stryker, equivalent FCS variant, and/or other fighting vehicles of comparable size and an additional 900 smaller vehicles. Combat sustainment was based on the requirement of 900 tons per day. By comparison, a modular heavy brigade weighs roughly 25,000 tons and the Future Combat System (FCS) brigade is projected to weigh 18,700 tons. Calculations assumed that a C-17 can carry three Stryker vehicles and, because slat armor must be removed in order for the vehicle to fit in a C-130, three C-130s were required to carry two Stryker vehicles. While many variables determine cargo load for airland, traditional airdrop, JPADS, and LAPES missions, a nominal capacity of 15 tons was used for all C-130 missions. The types of missions envisioned in this study drove an assumed JCA payload of 8 tons or 16,000 pounds. For C-17s, a 60-ton capacity was used for missions involving the transport of three Strykers and 50 tons was used for all other airland and airdrop missions.[1] All 34 C-17s used for combat employment were dedicated to Stryker transport. At three Strykers per mission and three missions per day, 34 C-17s lifted 306 Strykers in these calculations. Seven C-130s flying three missions per day moved the additional 14 Stryker vehicles and slat armor. No other airlift platforms used in the study can carry the Stryker vehicle.

While the Air Force has initially ordered only 24 JCA, this study assumes the Service will procure 70 airframes as required by the inter-Service Memorandum of Agreement. The analysis assumes the Army, which has placed an order for 54 thus far, will have 75 total airframes as currently projected. Calculations assume 100

1 The maximum C-17 capacity of 85 short tons is reduced when landing on semi-prepared or unprepared surfaces. Full-flap assault landings, which limit cargo loads to 50 tons, are required during combat operations on shorter runways of 3,500 feet.

percent of this JCA fleet of 145 aircraft are nominally available for BCT combat employment and sustainment. In cases where deployment of all JCA airframes is not feasible, the resulting shortfall would be offset in this planning model by other airframes such as the U-28/PC-12. Additionally, the maximum capacity of four sorties per day allows for flexibility when some JCA aircraft become unavailable during airlift operations. Calculations are based on current projections of 190 C-17s in the Air Force inventory, of which 38 aircraft or roughly 20 percent are made available under surge conditions for simultaneous BCT employment/sustainment (see Figure A.1). Of the projected Service-wide fleet size of 268 C-130s, which includes C-130E, C-130H, and C-130J, this study assumes 234 or 87 percent are made available for these surge conditions. While potentially important elements of the tactical airlift system, CV-22s, CSAR helicopters, and U-28/PC-12 aircraft are not included in these calculations as the routine availability of these platforms for BCT lift is in question. Organic Army rotary-wing assets and airframes from other Services are also not included. The inclusion of such assets would allow for some

Combat Sustainment (1 Brigade)

System	Capacity	Sortie rate	Total
25xJCA	8 tons/mission	2.0 missions/day	400 tons/day
15xC-130	15 tons/mission	2.0 missions/day	450 tons/day
1xC-17	50 tons/mission	1.0 mission/day	50 tons/day
41 aircraft		*81 missions/day*	*900 tons/day*

Combat Sustainment (4 Brigades)

System	Capacity	Sortie rate	Total
100xJCA	8 tons/mission	2.0 missions/day	1,600 tons/day
60xC-130	15 tons/mission	2.0 missions/day	1,800 tons/day
4xC-17	50 tons/mission	1.0 mission/day	200 tons/day
164 aircraft		*324 missions/day*	*3,600 tons/day*

Combat Employment (1 Brigade)

System	Capacity	Sortie rate	Total
45xJCA	8 tons/mission	2.0 missions/day	1,080 tons/day
174xC-130	15 tons/mission	2.0 missions/day	7,830 tons/day
34xC-17	60 tons/mission	3.0 mission/day	6,120 tons/day
253 aircraft		*759 missions/day*	*15,030 tons/day*

Figure A.1 Airlift calculations

reduction in the high numbers of airframes currently included in the calculations. Under the assumption that the enemy threat would not allow ground transport, truck convoys utilizing fighter escort is another option available under the hybrid "system of systems" approach not exercised in this air-only analysis.

All told, 41 aircraft would fly 81 daily missions to sustain one brigade by air while 164 aircraft would fly 324 sorties to provide sustainment for four brigades. A total of 253 aircraft flying 759 missions under surge conditions would be required to employ a single Stryker brigade in a 24-hour period. These calculations indicate that, while no more than one medium brigade can be employed by air at one time, a second Stryker brigade could be moved over a 48-hour period immediately following the first BCT using many of the same assets. Additionally, these combat employments could take place while simultaneously sustaining the other fielded brigades. Following the back-to-back employment of two Stryker brigades, analysis indicates it would be necessary to terminate surge conditions for a certain period of time, continue sustaining up to four brigades, and tend to other airlift demands which received less attention during the three-day combat employment period.

There are three final points with important implications relating to these calculations and assumptions. First, a key assumption rendering these results is that airfields in the rear area will exist in sufficient size and quantity to logistically support airlift operations involving over 250 aircraft under surge conditions. Second, if the JCA were to operate independently, 145 aircraft carrying 8 tons each during two missions per day could carry 2,320 tons daily, sustaining roughly 2.5 brigades at 900 tons daily. If the same 145 aircraft flew three missions per day during surge operations supporting the combat employment of a BCT, they could transport 3,480 tons, roughly 23 percent of the required 15,000 tons. Third, while strategic lift considerations are beyond the scope of this study, these assumptions are intended to allow for the simultaneous flow of intertheater airlift as well. The entire fleet of 111 C-5s, 80 percent of C-17s, and a projected fleet of 179 KC-X, which will likely have substantial cargo capacity, would be available for intertheater airlift during the combat employment effort. Additionally, 13 percent of the C-130s would be available for other intratheater demands during the same surge conditions. Under routine combat sustainment conditions, all but four C-17s and nearly 75 percent of C-130s are available for other purposes.

Selected Bibliography

Adams, T. (Winter 2001-2002), "Future Warfare and the Decline of Human Decisionmaking", *Parameters*, 57-71.

Air Force Basic Doctrine (2003), Air Force Doctrine Document 1 (Maxwell Air Force Base, Alabama: U.S. Air Force Doctrine Center).

Air Force Concept of Operations Development (2005), Air Force Instruction 10-2801 (Washington D.C.: Headquarters U.S. Air Force).

Air Force Glossary (2003), Air Force Doctrine Document 1-2 (Maxwell Air Force Base, Alabama: U.S. Air Force Doctrine Center, 9 July 1999, May 2003 update) <http://www.dtic.mil/doctrine/jel/Service_ pubs/afd1_2m 03.pdf>, accessed on 13 January 2007.

Air Force Strategic Plan 2006-2008 (2006), Washington D.C.: Headquarters U.S. Air Force, <http://www.airforcestrategynet.hq.af.mil/documents/AF% 20Strat%20Plan%20 Book%20FINAL%20PRINT%208-30.pdf>, accessed on 15 November 2007.

Air Mobility Operations (2006), Air Force Doctrine Document 2-6 (Maxwell Air Force Base, Alabama: U.S. Air Force Doctrine Center).

Air Support Operations Center Enabling Concept (1 June 2006), a briefing prepared by Air Combat Command (HQ ACC/A3YC), U.S. Air Force.

Air Warfare (2000), Air Force Doctrine Document (AFDD) 2-1 (Maxwell Air Force Base, Alabama: U.S. Air Force Doctrine Center).

Airspace Command and Control (AC2) (31 August 2006), Information Paper (Maxwell Air Force Base, Alabama: U.S. Air Force Doctrine Center).

Akam, R. (4 December 2006), correspondence with authors in response to survey.

Anderson, C. (1 September 2006), *Service Brief on the U.S. Army* (briefing at the Marine Corps War College, Marine Corps Base Quantico, Virginia).

—— (17 October 2006), interview by authors in Omaha, Nebraska.

Antal, J. (June 2002), "Killing Snakes: Lessons Learned from the Fighting in Afghanistan", *Army*, 14-15.

The Army in Joint Operations: The Army's Future Force Capstone Concept 2015-2024 (2005), TRADOC Pamphlet 525-3-0, Version 2.0 (Fort Monroe, Virginia: U.S. Army Training and Doctrine Command).

Army Transformation and the Impact on the Air Force (March 2006), Doctrine Watch (Maxwell Air Force Base, Alabama: U.S. Air Force Doctrine Center).

Army Transformation and the Impact on the Mobility Air Forces (November 2006), Doctrine Watch (Maxwell Air Force Base, Alabama: U.S. Air Force Doctrine Center).

Army Vision 2010: The Way Ahead (2006), Washington D.C.: Headquarters U.S. Army, <http://www.army.mil/2010/way_ahead.htm>, accessed on 13 November 2007.

The Army's Future Combat System (2006), Washington D.C.: Congressional Budget Office Study.

Atkinson, R. (2004), *In the Company of Soldiers* (New York, New York: Henry Holt and Company).

Baker, J. and Hamilton, L. (2006), *Iraq Study Group Report* (New York, New York: Vintage Books).

Baker, P. and Chandrasekaran, R. (27 March 2003), "Iraqi Militia, Elite Forces Roll South Into Fierce Attack by U.S. Warplanes", *The Washingon Post*, A1, <http://www.washingtonpost.com/ac2/wp-dyn/A34148-2003Mar26?language=printer>, accessed on 11 August 2008.

Barnes, J. (20 November 2006), "Army Gives Rumsfeld Doctrine a Rewrite", *Los Angeles Times*, <http://council.smallwarsjournal.com/showthread.php?t=1502>, accessed on 13 November 2007.

Binford, R. (19 October 2006), interview by authors in Omaha, Nebraska.

Bingham, P. (Spring 2002), "Seeking Synergy Joint Effects-Based Operations", *Joint Forces Quarterly*, <http://findarticles.com/p/articles/mi_m0KNN/is_2002_Spring/ai_91914143>, accessed on 13 November 2007.

Blaker, J. (July 1996), "The Owens Legacy: The Former Vice Chairman of the Chiefs of Staff Laid the Groundwork for a Revolution", *Armed Forces Journal International*, 20-22.

Blitzer, W. (11 December 2001), "What Will the Remaining Taliban and Al Qaeda Fighters Do?", *CNN.com*, <www.cnn.com/TRANSCRIPTS/0112/11/wbr.02.html>, accessed on 13 November 2007.

Bolkcom, C. (30 May 2006), *Military Airlift: C-17 Aircraft Program* (Washington D.C.: Congressional Research Service Report for Congress).

Bower, C. et al. (1993), *The New Calculus: Analyzing Airpower's Changing Role in Joint Theater Campaigns* (Santa Monica, California: RAND Corporation).

Burns, J. (10 April 2001), *Strategic Airlift: Our Achilles' Heel* (Carlisle Barracks, Pennsylvania: U.S. Army War College Thesis).

Burns, J.F. (28 May 2006), "Iran and Iraq to Join to Seal Border Against Insurgents", *New York Times*, <http://www.nytimes.com/2006/05/28/world/middleeast/28iraq.html>, accessed on 13 November 2007.

Butler, A. (October 2001), "The CONOPS with a Difference", *Air Force Magazine Online*, <http://www.afa.org/magazine/Oct2001/1001conops.asp>, accessed on 15 November 2007.

C-27J Spartan for the JCA Program, <http://www.c-27j.com>, accessed on 30 December 2007.

Canfield, W. (15 March 2006), *Air Mobility Command Doctrine Watch on Joint Cargo Aircraft* (Scott Air Force Base, Illinois: U.S. Air Force Air Mobility Command).

Cebrowski, A. (2005), Foreword in *Rethinking the Principles of War*, (ed.) A. McIvor (Annapolis, Maryland: Naval Institute Press), ix-xiii.

Clausewitz, C. (1976), *On War*, (eds) M. Howard and P. Paret (Princeton, New Jersey: Princeton University Press).

CMTC Adler's TTP for CSS, Tactics, Techniques, and Procedures (2002), Newsletter No. 02-1 (Fort Monroe, Virginia: Center for Army Lessons Learned, U.S. Army Training and Doctrine Command).

Colvin, R. (20 November 2006), "U.S. Says 100 Fighters a Month Enter Iraq From Syria", *Reuters AlertNet*, <http://www.alertnet.org/thenews/newsdesk/COL051504.htm>, accessed on 13 November 2007.

Commanding a Digital Brigade Combat Team, Tactics, Techniques, and Procedures (2001), Special Edition No. 01-21 (Fort Monroe, Virginia: Center for Army Lessons Learned, U.S. Army Training and Doctrine Command).

Committee Defense Review Report (December 2006), Washington D.C.: U.S. Congress House Armed Services Committee, <http://www.house.gov/hasc/CDR/CDRFinal.pdf>, accessed 9 January 2007.

Concept of Operations for Joint Cargo Aircraft (31 August 2007), Scott Air Force Base, Illinois: U.S. Air Force Air Mobility Command.

Conversino, M. (Winter 1997-1998), "The Changed Nature of Strategic Air Attack", *Parameters*, 28-41.

Cook, H. (July-August 2004), "Sea-Basing and the Maritime Pre-positioning Force (Future)", *Military Review*, <http://usacac.leavenworth.army.mil/cac/milreview/download/ English/JulAug04/cook.pdf>, accessed on 13 November 2007.

Cornella, A. (28 June 2005), "Testimony of the Chairman before the Senate Appropriations Subcommittee on Military Construction and Veterans Affairs", Washington D.C.: Overseas Basing Commission Testimony.

"Counter-Insurgency Urgency" (21 August 2006), *Aviation Week and Space Technology*, 36.

Counterair Operations (2002), Air Force Doctrine Document (AFDD) 2-1.1 (Maxwell Air Force Base, Alabama: U.S. Air Force Doctrine Center), <http://www.dtic.mil/doctrine/jel/Service_pubs/afdd2_1_1.pdf>, accessed on 13 January 2007.

Counterland Operations (2006), Air Force Doctrine Document 2-1.3 (Maxwell Air Force Base, Alabama: U.S. Air Force Doctrine Center), <http://www.dtic.mil/doctrine/jel/Service_pubs/afdd2 _1_3.pdf >, accessed on 13 January 2007.

Crowder, G. (1 April 2003), interview by authors at Langley Air Force Base, Virginia.

Curry, P. (17 October 2006), interview by authors in Denver, Colorado.

Czelusta, M. (2003), *Global Strike Task Force and Stryker Brigade Combat Team: Prospects for Integration in the Forcible Entry Mission* (Fort Leavenworth, Kansas: U.S. Army Command and Staff College Thesis).

Davis, A. (2005), *Back to the Basics: An Aviation Solution to Counterinsurgent Warfare*, Air Command and Staff College Wright Flyer Paper No. 23 (Montgomery, Alabama: Air University Press).

De Oliveira, M. (24 October 2006), correspondence with authors in response to survey.

Defense Science Board Task Force on Mobility (2005), Washington D.C.: Office of the Under Secretary of Defense for Acquisition, Technology, and Logistics.

Department of Defense Transformation Planning Guidance (2003), Washington D.C.: U.S. Government Printing Office, <http://www.oft.osd.mil/library/library_ files/document_129 _Transformation_ Planning_Guidance_April_2003_1.pdf>, accessed on 3 February 2007.

Deptula, D. (10 February 2007), "Letter to the Editor: Meeting Air Force's Needs, Today and Tomorrow", *New York Times*, A14.

Deptula, D. and Dahl, S. (July-August 2003), "Transforming Joint Air-Ground Operations for 21st Century Battlespace", *Field Artillery*, 21-25.

DoD Dictionary of Military and Associated Terms (2001), Joint Publication 1-02 (Norfolk, Virginia: U.S. Joint Forces Command, 12 April 2001, as amended through 8 August 2006), <http://www.dtic.mil/doctrine/jel/ doddict/>, accessed on 13 January 2007.

Douhet, G. (1983), *The Command of the Air* (Washington D.C.: U.S. Government Printing Office).

Dudney, R. (December 2006), "A Force for the Long Run", *Air Force Magazine* 89: 12, 2.

Egginton, J. (1994), *Ground Maneuver and Air Interdiction: A Matter of Mutual Support at the Operational Level of War* (Maxwell Air Force Base, Alabama: Air University Press).

Ehrhard, T. (1996), *Making the Connection: An Air Strategy Analysis Framework* (Maxwell Air Force Base, Alabama: Air University Press).

Fairchild, M. (1940), "The Aim in War", Lecture AF-3-L delivered at the Air Corps Tactical School, Maxwell Air Force Base, Alabama on 27 May 1940, Air Force Historical Research Agency File No. 248.2021A-3 (Maxwell Air Force Base, Alabama: Air Force Historical Research Agency).

Falvo, A. (2002), *Concept Offers Warfighters Battlefield Big Picture—Millennium Challenge 2002 will Test New Ways of Tying Together Information to Present as a Common Relevant Operational Picture (CROP)* (Norfolk, Virginia: U.S. Joint Forces Command), <http://www.jfcom.mil/about/experiments/mc02/crop.htm>, accessed on 13 November 2007.

Fastabend, D. (2005), "Transformation and Operational Art" in *Rethinking the Principles of War*, (ed.) A. McIvor (Annapolis, Maryland: Naval Institute Press), 155-166.

Feickert, A. (2006), *U.S Army's Modular Redesign: Issues for Congress* (Washington D.C.: Congressional Research Service Report for Congress).

Field Manual (FM) 3-0: Operations (2001), Fort Monroe, Virginia: U.S. Army Training and Doctrine Command.

Field Manual (FM) 100-5: Operations (1982), Washington D.C.: Headquarters Department of the Army.

Field Manual (FM) 1-114: Air Cavalry Squadron and Troop Operations (2001), Fort Monroe, Virginia: U.S. Army Training and Doctrine Command.

Forbes, I. (2005), "Future Warfare and the Principles of War" in *Rethinking the Principles of War*, (ed.) A. McIvor (Annapolis, Maryland: Naval Institute Press), 143-166.

Future Ground Combat Environment (October 2006), Fort Monroe, Virginia: U.S. Army Training and Doctrine Command.

Garamone, J. (7 May 2006), "Border Police Take on Mission of Securing Iraq's Frontier", *Defense Link*, <http://www.defenselink.mil/news/May2006/20060507_5048.html>, accessed on 13 November 2007.

Garretson, P. (Spring 2005), "High-Altitude Intercontinental Precision Airdrop: A Revolution in Mobility Affairs (Could AMC Learn from the B-2 Model?)", *Air Force Journal of Logistics* 29:1, 39-42.

Gates, R. (30 November 2006), *Advanced Policy Questions for Dr. Robert M. Gates, Nominee to be Secretary of Defense* (Washington D.C.: Preparatory Testimony for Senate Confirmation Hearings).

—— (19 December 2006), *Secretary of Defense Message to the Department of Defense Personnel* (Washington D.C.: ALDODACT 18/06).

Givens, R. (2002), *Turning the Vertical Flank, Air Power as a Maneuver Force in the Theater Campaign* (Maxwell Air Force Base, Alabama: Air University Press).

Global Persistent Attack CONOPS (July 2006), Langley Air Force Base, Virginia: Headquarters Air Combat Command.

Gordon, J. et al. (January-February 2007), "Air-Mechanization: An Expensive and Fragile Concept", *Military Review*, 63-73.

Gordon, M. (18 June 1983), "The Army's 'Air-Land Battle' Doctrine Worries Allies, Upsets the Air Force", *National Journal*, 4F.

—— (3 April 2003), "A Nation at War: Strategy; Goal of U.S.: Avoid a Siege", *The New York Times*, <http://query.nytimes.com/gst/fullpage.html?res=9802E3D91F39F93 0A35757C0A9659C8B63&n=Top/Reference/Times%20Topics/People/G/Gordon,%20Michael%20R>, accessed on 13 November 2007.

Gordon, M. and Trainor, B. (1995), *The General's War: The Inside Story of the Conflict in the Gulf* (Boston, Massachusetts: Little, Brown and Company).

Graham, B. (30 April 1999), "General Says U.S. Readiness is Ailing", *The Washington Post*, <http://www.washingtonpost.com/wp-srv/inatl/longterm/balkans/stories/airwar043099.htm>, accessed on 10 August 2008.

Grant, R. (September 2002), "The Airpower of Anaconda", *Air Force Magazine* 85: 9, 60-68.

—— (June 2004), "Marine Air in the Mainstream", *Air Force Magazine* 87:6, <http://www.afa.org/magazine/june2004/0604marine.html>, accessed on 6 February 2007.

Gray, C. (2005), "The American Way of War" in *Rethinking the Principles of War*, (ed.) A. McIvor (Annapolis, Maryland: Naval Institute Press), 13-40.

Gulf War Air Power Survey (1992), Volume 2 Draft Documents and Materials (Maxwell Air Force Base, Alabama: Air Force Historical Research Agency Document No. TFM-72).

Hallion, R. (1992), *Storm Over Iraq: Air Power and the Gulf War* (Washington D.C.: Smithsonian Institute Press).

—— (1994), "The Future of Air Power" in *The War in the Air 1914-1994*, (ed.) A. Stephens (Royal Australiam Air Force Base, Fairbairn, Australia: Air Power Studies Center).

Hammond, G. (2005), "The U.S. Air Force and the American Way of War" in *Rethinking the Principles of War*, (ed.) A. McIvor (Annapolis, Maryland: Naval Institute Press), 109-126.

Hansen, R. (2002), *The Effectiveness of the A-10 on the Battlefield of 2010* (Fort Leavenworth, Kansas: U.S. Army Command and Staff College Thesis).

Haven, P. (18 March 2002), "As the fog of war lifts, Operation Anaconda comes into sharper focus", *The Brownsville Herald Online Edition*, <http://www.brownsvilleherald. com/onset?id=745&template=article.html>, accessed on 8 November 2007.

Hawley, R. et al. (September 2002), "Enhancing USAF's Pacific Posture: How the Air Force Can Transform to Support a New Joint Warfighting Architecture", *Armed Forces Journal International*, 54.

Hinman, E. (2007), *Toward AirLandBattle21: Transformational Concepts for Integrating 21st Century Air and Ground Forces* (Marine Corps Base Quantico, Virginia: Marine Corps War College Thesis).

—— (First Quarter 2008), "Counterair and Counterland Concepts for the 21st Century", *Joint Force Quarterly*, 85-91.

Hirschberg, M. (1999), "On the Vertical Horizon: Bell Designs Are Accelerating at Full Tilt", *Vertiflight Articles for 1999* (Alexandria, Virginia: AHS International, The Vertical Flight Society), <http://www.vtol.org/vertiflite/vindex99.htm>, accessed on 16 November 2007.

Horne, A. (1978), *Savage War of Peace: Algeria 1954-1962* (New York, New York: Viking Press).

Hosmer, S. (1996), *Psychological Effects of U.S. Air Operations in Four Wars, 1941-1991* (Santa Monica, California: RAND Corporation).

Hymes, C. (24 October 2006), correspondence with authors in response to survey.

Jinnette, J. (2003), *Employing an Air Maneuver Force: Battlefield Air Operations with Surface Maneuver in a Joint Campaign* (Maxwell Air Force Base, Alabama: U.S. Air Command and Staff College Thesis).

Johnson, D. (26 September 2006), correspondence with the authors.

Joint Doctrine for Barriers, Obstacles, and Mine Warfare (2007), Joint Publication 3-15 (Norfolk, Virginia: U.S. Joint Forces Command), <http://www.fas.org/irp/doddir/dod/jp3_15.pdf>, accessed on 8 November 2007.

Joint Doctrine for Forcible Entry Operations (2001), Joint Publication 3-18 (Norfolk, Virginia: U.S. Joint Forces Command).

Joint Force Land Component Commander Handbook (2001), *Field Manual (FM)* 3-31 and MCWP 3-40.7 (Fort Monroe, Virginia: U.S. Army Training and Command).

Joint Readiness Training Center Heavy Team Handbook: Integration with the Light Brigade Combat Team (2002), Handbook No. 02-14 (Fort Monroe, Virginia: Center for Army Lessons Learned, U.S. Army Training and Doctrine Command).

Jordan, B. (30 October 2006), "JCA critics helped by RAND draft report", *Air Force Times* 67:15, 23.

Kaplan, L. (25 March 2002), "Troop Movement: How the Army ditched the Powell Doctrine", *The New Republic*, <http://www.tnr.com/>, accessed on 11 August 2008.

Kill Box: Multi-Service Tactics, Techniques, and Procedures for Kill Box Employment (2005), Norfolk, Virginia: Army, Marine Corps, Navy, Air Force Air Land Sea Application Center.

Klaus, J. (29 April 2005), *Strategic Mobility Innovation: Options and Oversight Issues* (Washington D.C.: Congressional Research Service Report for Congress).

Lambeth, B. (2000), *The Transformation of American Air Power* (Ithaca, New York: Cornell University Press).

—— (Spring 2002), "Lessons from the War in Kosovo", *Joint Forces Quarterly*, <http://findarticles.com/p/articles/mi_m0KNN/is_2002_Spring/ai_91914138?tag=artBody;col1>, accessed on 11 August 2008.

—— (2005), *Air Power Against Terror: America's Conduct of Operation Enduring Freedom* (Santa Monica, California: RAND Corporation).

Leonhard, R. (1991), *The Art of Maneuver: Maneuver-Warfare Theory and AirLand Battle* (Novato, California: Presidio Press).

—— (2005), "From Operational Art to Grand Strategy" in *Rethinking the Principles of War*, (ed.) A. McIvor (Annapolis, Maryland: Naval Institute Press), 208-222.

London, H. (1984), *Military Doctrine and the American Character: Reflections of AirLand Battle* (New York: National Strategy Information Center).

Long, A. (2006), *On Other War Lessons from Five Decades of RAND Counterinsurgency Research* (Washington D.C.: RAND Corporation).

Long, J. (1991), *The Evolution of U.S. Army Doctrine: From Active Defense to AirLand Battle and Beyond* (Alexandria, Virginia: Defense Technical Information Center).

"Low Altitude Parachute Extraction System" (2008), *GlobalSecurity.org*, <http://www.globalsecurity.org/military/systems/aircraft/systems/lapes.htm>, accessed on 10 January 2008.

Luck, G. (November 2006), "Insights on Joint Operations: The Art and Science, Best Practices, The Move toward Coherently Integrated Joint, Interagency, and Multinational Operations" in *A Common Perspective: U.S. Joint Forces Command Joint Warfighting Center Joint Doctrine Group's Newsletter* 14: 2, 1-37.

Mao Tse Tung (1975), "Problems of Strategy in China's Revolutionary War" in *Selected Works of Mao Tse-Tung, Volume I* (Peking, China: Foreign Language Press).

Marine Corps Operations (2001), Marine Corps Doctrinal Publication (MCDP) 1-0 (Washington D.C.: Department of the Navy, Headquarters United States Marine Corps).

Matsumura, J. (2000), *Lightning Over Water: Sharpening America's Light Force for Rapid-Reaction Missions* (Washington D.C.: RAND Corporation).

—— (2003), *Future Combat Systems Program* (Washington D.C.: RAND Corporation).

Mazzetti, M. (2 April 2003), "The Battle for Baghdad begins: Marine and Army Units Advance through Republican Guard Defenses, Web Exclusive, Conflict With Iraq: Front Line Report", *U.S. News and World Report*, <http://www.usnews.com/usnews/news/ iraq/articles/battle030402.htm>, accessed on 14 November 2007.

McCaffrey, B. (January 2003), "The Challenges to U.S. Security Armored Formations Will Remain Essential in Future Combat", *Armed Forces Journal*, <http://www.armedforcesjournal.com/2003/01/>, accessed on 11 August 2008.

McCaffrey, T. (2002), *What Happened to BAI? Army and Air Force Battlefield Doctrine Development From Pre-Desert Storm to 2001* (Maxwell Air Force Base, Alabama: School Of Advanced Airpower Studies Thesis).

McCoy, J. (September/October 2001), "Configuring Airdrop Packages for the IBCT", *Army Logistician*, 12-14.

McElroy, R. (September-October 2002), "Afghanistan Fire Support for Operation Anaconda", *Field Artillery*, 5-9.

McGowan, L. (August 2006), "Airdrop System Resupplies Ground Troops", *Defense Link: U.S. Department of Defense Transformation*, <http://www.defenselink.mil/transformation/articles/2006-08/ta081606a.html>, accessed on 14 November 2007.

McMahon, M. (2001), "Adaptive Transformation Model: A Branch to the Army Transformation Campaign Plan" in *Army Transformation: A View from the U.S. Army War College*, (ed.) W. Murray (Carlisle, Pennsylvania: Strategic Studies Institute), 195-223.

Mead, J. (24 October 2006), correspondence with authors in response to survey.

Meilinger, P. (1997), *Paths of Heaven: The Evolution of Airpower Theory* (Maxwell Air Force Base, Alabama: Air University Press).

—— (2006), *Dog Days for the Air Force: What's Wrong and How Can it be Fixed*, unpublished manuscript.

Melillo, M. (Autumn 2006), "Outfitting a Big-War Military with Small-War Capabilities", *Parameters* XXXVI:3, 22-35.

"Military Worth Analysis of New Concept Weapons" (2007), *Defense Tech Briefs*, <http://www.defensetechbriefs.com/ content/view/1081/34/1/1>, accessed on 8 November 2007, 1.

"MMIST Sherpa" (2007), *Mist Mobility Integrated Systems Technology (MMIST) Inc.*, <http://www.mmist.ca/Sherpa.asp>, accessed on 12 January 2007.

"Modular Force Update" (2 November 2006), *SOPS-T Brief* (Fort Leavenworth, Kansas: U.S. Army Combined Arms Center, Current Force Integration Directorate).

Moore, C. (January 2003), *Sea Basing: Operational Independence for a New Century* (Annapolis, Maryland: U.S. Naval Institute).

Mueller, K. (Spring 1998), "Denial, Punishment, and the Future of Air Power", *Security Studies* 7:3, 182-228.

Murray, W. (ed.) (2001), *Army Transformation: A View from the U.S. Army War College* (Carlisle Barracks, Pennsylvania: Strategic Studies Institute).

Murray, W. and Scales, R. (2003), *The Iraq War: A Military History* (London: The Belknap Press of Harvard University Press).

Nalty, B. (1986), *Air Power and the Fight for Khe Sanh* (Washington D.C.: Office of Air Force History, Headquarters U.S. Air Force).

The National Defense Strategy of the United States of America (2008), Washington D.C.: U.S. Government Printing Office.

The National Military Strategy of the United States of America: A Strategy for Today; A Vision for Tomorrow (2004), Washington D.C.: U.S. Government Printing Office.

The National Security Strategy of the United States of America (2006), Washington D.C.: U.S. Government Printing Office.

Nelms, D. (October 2005), "The New Army", *Rotor and Wing*, 1.

Nurkin, T. (18 October 2006), "JSAS White Paper - UAS: unmanned, unbound", *Jane's*, <http://www.janes.com/defence/news/jsas/jsas061018_1_n.shtml>, accessed on 14 November 2007.

Operations and Organizations (2007), Air Force Doctrine Document 2 (Maxwell Air Force Base, Alabama: U.S. Air Force Doctrine Center), <http://www.fas.org/irp/doddir/ usaf/afdd2.pdf>, accessed on 26 February 2007.

Options for Restructuring the Army (May 2005), Washington D.C.: Congressional Budget Office Study.

Overy, R. (1980), *The Air War, 1939-1945* (New York, New York: Stein and Day Publishers).

"Pakistan Willing to Seal Off Border With Afghanistan" (6 November 2006), *Voice of America News*, <http://www.voanews.com/english/archive/2006-11/2006-11-06-voa3.cfm?CFID= 234207522&CFTOKEN=50912766>, accessed on 14 November 2007.

Palmer, P. et al. (14 April 1998), *The Battle of Khafji: An Assessment of Airpower* (Maxwell Air Force Base, Alabama: Air War College).

Pape, R. (1996), *Bombing to Win: Air Power and Coercion in War* (Ithaca, New York: Cornell University Press).

Quadrennial Defense Review Report (6 February 2006), Washington D.C.: U.S. Government Printing Office.

Record, J. (16 October 2006), "Why America Keeps Losing 'Small Wars'", *The Baltimore Sun*, <http://www.baltimoresun.com/news/opinion/oped/bal-op.smallwars15oct15,0,5948046.story?coll=bal-oped-headlines>, accessed on 14 November 2007.

Reynolds, R. (1995), *Heart of the Storm: The Genesis of the Air Campaign against Iraq* (Maxwell Air Force Base, Alabama: Air University Press).

Richburg, K. and Glasser, S. (27 March 2003), "Iraqi Tanks Try to Break Out of Basra, British Troops Bombard City", *The Washington Post*, <http://www.washington post.com/wp-dyn/nation/nationalsecurity/abroad/iraq/field/glasserSusan>, accessed on 14 November 2007.

Risher, P. (2005), "Teaching, Learning, and Leading: The New Mandate" in *Rethinking the Principles of War*, (ed.) A. McIvor (Annapolis, Maryland: Naval Institute Press), 244-260.

Rolfsen, B. (6 November 2006), "CV-22 Osprey: The New Warbird", *Air Force Times* 67:16, 16-18, 20.

—— (12 February 2007), "Spec ops chief reveals plan to diversify fleet", *Air Force Times* 67: 30, 18.

Romjue, J. (1993), *Prepare the Army for War: A Historical Overview of the Army Training and Doctrine Command, 1973-1993* (Fort Monroe, Virginia: U.S. Army Doctrine and Training Command Office of the Command Historian), <http://cgsc.cdmhost.com/cdm4/document.php?CISOROOT=/p4013coll11&CISOPTR=713&REC=1&CISOSHOW=705>, accessed on 7 August 2008.

Roosevelt, A. (9 May 2005), "Air Force Takes On Stryker Vehicles", *Defense Daily* 226:27, 1.

—— (5 Jun 2006), "Army Concerned If Congress Takes Funds From Joint Cargo Aircraft Program", *Defense Daily* 230: 45, 1.

Rumsfeld, D. (12 May 2005), *U.S. Department of Defense News Briefing* (Washington D.C.: U.S. Department of Defense), <http://www.defenselink.mil/transcripts/2005/tr20050512-secdef2761.html>, accessed on 14 November 2007.

Rumsfeld, D. and Myers, R. (3 April 2003), *Defense Department Operational Update Briefing* (Washington D.C.: Federal News Service), <http://web.lexis-nexis.com/universe/document>, accessed on 14 November 2007.

Scales, R. (2005), "The Second Learning Revolution" in *Rethinking the Principles of War*, (ed.) A. McIvor (Annapolis, Maryland: Naval Institute Press), 41-57.

—— (July 2006), "Clausewitz and World War IV", *Armed Forces Journal*, <http://www. armedforcesjournal.com/2006/07/1866019>, accessed on 14 November 2007.

Scarborough, R. (27 April 2003), "Rulers of the air", *The Washington Times*, 6, <https://www. washtimes.com/national/20030427-10772613.htm>, accessed on 8 November 2007.

Schoomaker, P. (27 June 2006), *Statement by General Peter J. Schoomaker, Chief of Staff of the Army, Before the Committee on Armed Services, United States House of Representatives, Second Session, 109th Congress, on the Army's Reset Strategy and Plan for Funding Reset Requirements*.

Scully, M. (11 July 2006), "Review of Army combat program challenges need for lighter vehicles", *Congress Daily*, <http://www.govexec.com/story_page.cfm?articleid= 34517&sod-21>, accessed on 14 November 2007.

Sea Basing Perspective Within the Triad of Sea Power 21 (November 2004), Sea-Basing White Paper (Bremerton, Washington: AMI International), <http://www.amiinter.com/SeaBasingProgramNov2004.pdf>, accessed on 14 November 2007.

Search and Cordon: Multi-Service Tactics, Techniques, and Procedures for Search and Cordon Operations (2006), Norfolk, Virginia: Army, Marine Corps, Navy, Air Force Air Land Sea Application Center.

Serfass, S. (April 2004), *C-17 Low Altitude Parachute Extraction System Employment* (Fort Dix, New Jersey: U.S. Air Force Mobility Weapons School).

Shannon, E. and McGirk, T. (24 November 2006), "Iran and Syria Helping Hezbollah Rearm", *Time*, <http://www.time.com/time/world/article/0,8599,1562890,00.html>, accessed on 14 November 2007.

Shinseki, E. (October 2000), "The Army Transformation: A Historic Opportunity", *Army*.

Slessor, J. (1936), *Airpower and Armies* (London: Oxford University Press).

Smucker, P. (4 March 2002), "How bin Laden Got Away", *The Christian Science Monitor*, <http://www.csmonitor.com/2002/0304/p01s03-wosc.html>, accessed on 14 November 2007.

Staniland, P. (Winter 2005-2006), "Defeating Transnational Insurgencies: The Best Office Is a Good Fence", *Washington Quarterly* 29:1, 21-40.

Steele, D. (April 2002), "Operation Anaconda, Taking the Fight to the Enemy in Afghanistan", *Army*, 19-20.

St-Pierre, H. (6 November 2006), interview by authors at Carlisle Barracks in Pennsylvania.

Sun Tzu (1963), *The Art of War*, translated by S. Griffith (London: Oxford University Press).

—— (1993), *The Art of War: The New Translation*, research and reinterpretation by J. Huang (New York, New York: Quill William Morrow).

Svetecz, T. (2001), *Global Strategic Task Force: A Strategic Renaissance* (Maxwell Air Force Base, Alabama, Air Command and Staff College Thesis), <https://research.au.af.mil/papers/ay2001/acsc/01-116.pdf >, accessed 13 January 2007.

Tactical Air Control Party Enabling Concept (2006), HQ ACC/A3YC Draft White Paper (Langley Air Force Base, Virginia: U.S. Air Force Air Combat Command).

Talaber, A. (May 2005), *Options for Restructuring the Army* (Washington D.C.: Congressional Budget Office).

Talbott, J. (1980), *The War Without a Name: France in Algeria, 1954-1962* (New York, New York: Alfred Knopf).

Tirpak, J. (October 2004), "The Airlift Gap", *Air Force Magazine* 87:10, <http://www.afa.org/magazine/Oct2004/1004gap.asp>, accessed on 14 November 2007.

—— (March 2006), "Rising Risk in Air Mobility", *Air Force Magazine* 89:3, <http://www.afa.org/magazine/march2006/0306mobility.asp>, accessed on 14 November 2007.

—— (December 2007), "Crunch Time for Air Mobility", *Air Force Magazine* 90:12, 29-33.

Toler, L. (Winter 2003), "Distribution Management: What's Really Different", *Quartermaster Commentary*, <http://www.quartermaster.army.mil/oqmg/Professional_Bulletin/2003/Winter03/QM_Commentary_Distribution_Management_What's_Really_ Different.htm>, accessed on 12 January 2007.

Transforming Military Operational Capabilities (27 April 2001), Transformation Study Report (Washington D.C.: Office of the Secretary of Defense), <http://stinet.dtic.mil/oai/oai? verb=getRecord&metadataPrefix=html&identifier=AD A392289>, accessed on 14 November 2007.

Trowbridge, G. (28 April 2003), "Air Power Paves Way, Leaders recount success, lessons learned from working with ground units", *Air Force Times*, 8.

The United States Army 2004 Posture Statement (2004), Washington D.C.: Headquarters U.S. Army, <http://www.army.mil/aps/04/core.html>, accessed on 14 November 2007.

The United States Army Concept for Tactical Maneuver for the Future Modular Force: 2015-2024 (2006), TRADOC Pamphlet 525-3-2, Version 1.0 (Fort Monroe, Virginia: U.S. Army Training and Doctrine Command), <http://www-tradoc.army.mil/tpubs/ pams/p525-3-2.pdf>, accessed on 31 January 2007.

The United States Army Operating Concept for Operational Maneuver: 2015-2024 (2006), TRADOC Pamphlet 525-3-1, Version 1.0 (Fort Monroe, Virginia: U.S. Army Training and Doctrine Command).

The United States Strategic Bombing Surveys (1987), Maxwell Air Force Base, Alabama: Air University Press.

Unmanned Aircraft (UA) and Airspace Control in the Combat Zone (December 2005), Doctrine Watch (Maxwell Air Force Base, Alabama: U.S. Air Force Doctrine Center).

Van Creveld, M. (1994), *Air Power and Maneuver Warfare* (Maxwell Air Force Base, Alabama: Air University Press).

Vick, A. et al. (2002), *The Stryker Brigade Combat Team: Rethinking Strategic Responsiveness and Assessing Deployment Options* (Arlington, Virginia: RAND Corporation).

Warden, J. (1992), "Employing Air Power in the Twenty-first century" in *The Future of Air Power in the Aftermath of the Gulf War*, (eds) R. Shultz and R. Pfaltzgraff (Maxwell Air Force Base, Alabama: Air University Press).

—— (Spring 1995), "The Enemy as a System", *Airpower Journal* 9:2, 40-55.

—— (1998), *The Air Campaign: Planning for Combat* (New York, New York: toExcel).

Warfighting (1997), Marine Corps Doctrinal Publication (MCDP) 1 (Washington D.C.: Department of the Navy, Headquarters United States Marine Corps), <http://www.dtic.mil/ doctrine/jel/Service_pubs/mcdp1.pdf>, accessed on 27 January 2007.

Wathen, A. (29 September 2006), *JCA Déjà Vu: The Distributed Battlefield Logistics War—All Over Again* (Maxwell Air Force Base, Alabama: Air University Press), <https://research.maxwell.af.mil/Publication.aspx?PubCat=Airpower+Research+Institute+Quick+Looks&PubYr=0>, accessed on 14 November 2007.

Way Ahead for the Convergence of the Army Future Cargo Aircraft and Air Force Light Cargo Aircraft Programs (16 June 2006), Memorandum of Agreement between the Department of the Army and the Department of the Air Force, Volume 8.5.

Westenhoff, C. (October 1990), *Military Air Power: The CADRE Digest of Air Power Opinions and Thoughts* (Maxwell Air Force Base, Alabama: Air University Press).

Williams, K. (1 August 2006), *Air Force Doctrine Update* (briefing by U.S. Air Force Doctrine Center at the Marine Corps War College, Marine Corps Base Quantico, Virginia).

Wilson, P. (1974), *Defense of Hue and Quangtri City: The 1972 NVN Invasion of MR-1* (Alexandria, Virginia: Center for Naval Analyses).

Wilson, P. et al. (Winter 2003-2004), "An Alternative Future Force: Building a Better Army", *Parameters*, <carlisle-www.army.mil/ usawc/Parameters/03winter/wilson.pdf>, accessed on 14 November 2007.

Winnefeld, J. and Johnson, D. (1993), *Joint Air Operations Pursuit of Unity in Command and Control 1942-1991* (Annapolis, Maryland: Naval Institute Press).

Worden, M. (1998), *Rise of the Fighter Generals: The Problem of Air Force Leadership, 1945-1982* (Maxwell Air Force Base, Alabama: Air University Press).

Zastrow, R. (13 December 2006), *USAF Employment ... A Strategic Perspective* (briefing at the Army War College, Carlisle Barracks, Pennsylvania).

Zeeck, K. (2006), *Army Modular Forces* (briefing by U.S. Air Force, Doctrine Application Directorate, U.S. Air Force Doctrine Center at Maxwell Air Force Base, Alabama).

Zielinski, P. (6 November 2006), interview by author at Carlisle Barracks in Pennsylvania.

Index